NUMERICAL MATHEMATICS

EXERCISES IN COMPUTING WITH A DESK CALCULATOR

INTRODUCTORY MONOGRAPHS IN MATHEMATICS

General Editor
(the late) A. J. Moakes

NUMERICAL MATHEMATICS

Exercises in computing with a desk calculator

A. J. MOAKES, M.A., F.I.M.A.

Third Edition
completed by H. Neill

MACMILLAN

First published 1963
Second edition 1965
Third edition 1973

Published by
THE MACMILLAN PRESS LTD
London and Basingstoke
Associated companies in New York,
Melbourne, Dublin, Johannesburg and Madras

SBN 333 14686 7

Reproduced and printed by photolithography and bound in
Great Britain at The Pitman Press, Bath

PREFACE TO THIRD EDITION

SINCE the first edition was published, numerical work with machines has come to play a regular part in education whether in mathematics or as a step to computer studies. This fact has been recognised in some syllabuses, notably that of the project for Mathematics in Education and Industry (M.E.I.). Since this text has been used in preparation for such examinations, it has been expanded to include suitable revision questions. The author wishes to thank the Oxford and Cambridge Board for leave to reprint questions from the M.E.I. A Level and Special Papers, and also to the Examinations Council of London University for questions from that University's Teachers' Certificate.

The bibliography and other references have been brought up to date. Attention is drawn to the additional note on Tables (Appendix IV). The computer, while it has made both desk-machines and tables into subsidiary aids, has led us to demand more exacting standards from them; and in the field of function tables especially a notable change in usage is currently taking place.

A. J. M.

PREFACE TO FIRST EDITION

THE view is gaining ground in the educational world that to free the learner (after a certain point) from laborious calculation is to help him to grasp and develop the underlying ideas. These exercises are therefore devised to *use the machine so as to stimulate mathematical thought*. This aim has been pursued from the beginning but only becomes fully apparent after the full range of machine operations has been mastered.

This collection of exercises, along with the necessary minimum of explanation, was devised for the use of sixth-formers. It should be helpful to those teachers who want their pupils to be introduced to machines but who are not clear how to set about it.

In the first five chapters the technique of handling a machine is developed stage by stage through progressive examples. For each stage it is necessary first to learn certain manual operations, preferably by seeing them demonstrated and immediately imitating them. For a course in being, one pupil can demonstrate to another — to their mutual advantage. A teacher wishing to initiate such a course can acquire the skills with the aid of (*a*) a full demonstration by the maker's representative, followed by (*b*) thorough practice, with the maker's handbook for reference.

The author wishes to express gratitude, for their encouragement and help, to Mr. M. Shoenberg (educational adviser of Olympia Business Machines, who make the Brunsviga), to Mr. F. W. Russell (education manager, Monroe Calculating Machines), to Dr. J. Crank (Head of Mathematics Department, Brunel College), to Mr. M. N. Horsman of Brunel College, as well as to Mr. D. Grisewood of Macmillan & Co. Ltd. and to Mr. M. Bridger of Leicester College of Technology.

Acknowledgement must also be made to the author's pupils at St. Paul's School, who have greeted machines with enthusiasm, treated them with respect, and made shrewd comments on the exercises.

Standing orders for pupils doing the course are reproduced as an appendix, together with some hints which may prove useful to teachers who wish to use a machine occasionally for junior class-teaching.

A. J. M.

CONTENTS

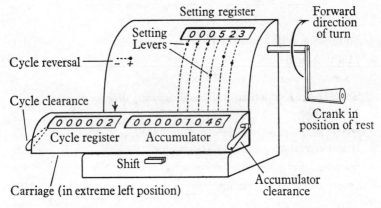

Basic form of desk machine (lever-setting type)

INTRODUCTION

The Desk Calculating Machine

MACHINES operated entirely by hand invariably incorporate the following registers, where the numbers which are involved in the calculation are exhibited:

(1) The input or *setting register*.

A number is set on the machine by the pressing of keys or the setting of levers; and the setting register indicates this number, as a check that (*a*) its digits have been set correctly and (*b*) it is correctly located.

(2) The *cycle register* or multiplier register.

The processing of the input number is done entirely by turns of the operating handle, e.g. a single forward turn of this handle causes the input number N to be added in to the accumulator once (while still remaining set for further use). This turn is recorded as a unit figure in the cycle register. Seven such turns would show the number 7 and give $7N$ in the accumulator: this fact explains the name 'multiplier register' often employed.

(3) The *accumulator*, already mentioned, can also be called the product register as the above example explains; but the former term explains more clearly its usual function, viz.: to accumulate (so long as no action is taken to clear it) all numbers which, singly or as multiples, are brought into it by the action of the operating handle.

The Operation of the Machine

This is described in general terms which apply to any machine except where special reference is made to a particular model.

The basic function of the *operating handle* has been described; but its operation would be very limited without the *shift lever*, a device which by lateral movement of the accumulator causes the numbers to be fed into it at any position. E.g. with the input one place to the left (relatively) the operation described in (2) would feed in not $7N$ but $70N$.

It must be possible to clear any register, and we have therefore three separate *clearing levers*, and sometimes a *total clearance lever* which clears all registers in one motion. (It is however possible, e.g. in the *Brunsviga 20*, to lock the left-hand half of the accumulator as a storage, and the clearance will then not operate on this half).

Another valuable facility on many machines is a *back-transfer device* by which a result can, while being cleared from the accumulator, also be automatically *set*. By adjustment of the shift, this setting can either be done complete or with the final digits omitted. Generally, all but the smallest machines now have this back-transfer facility. The possession of a machine with such a device eliminates one source of error (viz. resetting by hand) as well as speeding up certain programmes of calculation.

A list of firms marketing hand-machines in the U.K. is given on p. 61, with some notes on choosing a machine for teaching purposes.

Electrically driven machines are basically the same as the hand-machine in their mechanical construction, but fall into three classes according to their degree of automatic operation:

(1) Semi-automatic machines, in which division is automatic but not multiplication;

(2) machines in which the multiplicand is set on the keyboard and the multiplication process is then carried out digit by digit; and

(3) fully automatic machines, in which both division and multiplication can be preset.

The possibility of jamming a machine is increased by automatic operation. It usually occurs because a new process is initiated before the machine has completed the previous one.

Electronic desk machines have the double advantage of being silent and of being unjammable. It does, however, make considerable demands on the skill of the operator to get results of the full accuracy of which such a machine is capable. This skill in the suitable setting of numbers (by adjusting powers of ten) can be acquired on a basic type of hand machine.

Note for the Learner

To the learner: the course is planned as follows:

Before each of the early chapters you must have certain machine operations shown to you on your own type of machine. These are listed in the notes at the appropriate stages. Generally, anyone who has successfully completed the exercises can act as demonstrator for you.

Repeat the manual operations until you understand them and do not let the demonstrator go until *you* are satisfied. You should then *do the exercise without help* even if it means trying it a number of times.

There are no 'answers at the back'. In every case you will be able to verify your own numerical answers by the checking method indicated.

Keep a special 'maths practical book' for this work. Record all your results, both intermediate and final; together with your method if any choice was given, and your check. A clear and systematic layout is of great importance, and in the more complex examples a specific form of layout is given for you to follow.

Where there are verbal questions, answer them in the book. Add any useful comments of your own *at the time*: this will increase the value of the record. Solution notes are given on pp. 62 onwards, to questions which seem to demand them. The existence of such a note is indicated in each case thus: [Solution note.

In professional numerical work *no* answer is given without an indication of its acceptability — by rounding off figures or otherwise. In this course you will be gradually subjecting your results to criticism; but you should from the beginning avoid copying down figures from the machine 'just because they are there'.

The book consists of 10 chapters. An exercise numbered 8.6 is the 6th exercise in the 8th chapter. The early explanatory notes precede the exercises to which they refer, but from chapter 4 onwards the left-hand pages are reserved for notes and the right-hand for the exercises.

Considerable cross-reference is used where techniques arise again in a new context, but for additional guidance you will find on each page a short summary of the contents, and at the end of the book is a short index of terms, giving in each case the page where the term first appears.

SUMMATION AND THE DIFFERENCE TABLE

Learn how
  (i) to clear all registers;
 (ii) to return the carriage to first position;
(iii) to set decimal point markers for addition/subtraction;
 (iv) to set up a number on the levers or keyboard, and to verify that it is correctly set;
  (v) to add this number into the accumulator;
 (vi) to notice the indication given that an addition has been carried out;
(vii) to clear the keyboard only, ready to set and add a further number.

1.1. Add **592** and **89**, in this order and note the result. Then add in the reverse order.

1.2. Add **8·649** and **14·7** in this order. Note.
Add 14·700 and 8·649 in this order.
If the results do not agree, probably your decimal point setting was not correct: the decimal point marker, set for the largest number of decimal places (3 here), applies to all numbers set.

1.3. Add each row of the following table, noting the results and then adding them to give a grand total.

27·94	6·79	18·312	43·0 =sum of row
31·02	28·4	43·89	17·4 = ,, ,, ,,
6·75	5·81	1·08	25·02 = ,, ,, ,,
			total

Now repeat the summation, but add first the columns, and then sum the *column* totals to a grand total.

What is the significance of the number which appears in the cycle register at any stage of the calculation 1.3?

For 1.4. Learn

  (i) how to subtract a number from the accumulator;

 (ii) what happens if the number subtracted is greater than that in the accumulator;

(iii) when it is possible to disregard the appearance of such results (see note on *complements* below).

(iv) how to restate a number obtained as a complement, in a form with a negative sign. (Calculate mentally; and add back, as a check, to give zero.)

1.4. Add the rows and columns of the table below in the same way as in 1.3, and note all results *just as they appear on the machine*. Form the grand totals as before.

72·94	−6·97	18·321	−40·3
−13·02	28·4	43·89	−17·4
6·75	−5·81	−1·08	25·02

1.5. Convert the result of adding the 4th column of 1.4 into a negatively signed number.

1.6. Set a number which includes a decimal point. Add it into the accumulator 11 times.

(You have thus learnt how to multiply by an integer).

Would it be possible to get this result by only two turns of the handle instead of 11?

A note on complements

If you subtract unity from an 8-digit (cleared) register it shows 99,999,999; while $0 - N$ shows $(10^8 - N) = N'$ say. N' is called the *complement* of N, since they make up together a standard total (cf. complementary angle in geometry).

An all-machine method of finding N from N' (standing in the accumulator) is: set N', and subtract it twice from the accumulator. The first subtraction gives zero, and checks the setting: the second gives N, together with some irrelevant 9's in the extra digit-places of the left half of the accumulator. (Try this.)

1.7.

A difference table, for a function at equal intervals.

Serial No.	Value	1st Difference	2nd Difference	3rd Difference
0	− 10			
		+6		
1	− 4		−2	
		+4		0
2	0		−2	
		+2		0
3	+ 2		−2	
		0		
4	+ 2			

Above is an example of a 'difference table' for a series of values. Whenever such values are derived by putting successive integral values for n in an algebraic (polynomial) formula, the differences become equal all the way down one of the columns, and zero in the next, as seen here.

Set out a difference table for nos. $0, 1, 8, 27, 64, 125, 216$. In the column in which the differences become equal, put an extra such difference at the foot. Use it to construct an eighth number of the sequence given. An example of this procedure is shown in the note for Ex. 4.4. If you can also fit such an algebraic formula $f(n)$ to the given values and calculate $f(7)$, then you will have checked your work. Otherwise, the check for differences is to add back, e.g. in this table the first differences are checked by starting at the top of the 'values' column, and then adding successively 6, 4, 2, 0, getting at each stage the next number down in the value column. (It may save time in simple cases to subtract mentally to *make* the table, and only use the machine to *check* it.)

1.8. Construct a cubic polynomial $an^3 + bn^2 + cn + d$ in which a, b, c, d are chosen integers (positive or negative). Calculate the values of $f(n)$ for values of n from zero to 5 i.e. **tabulate f(n) for n = 0(1)5;** the bracket denotes the interval between successive values of n. Form the difference table and verify that the third differences are equal. Show algebraically that such equal differences must have the value $3 \times 2 \times a = 6a$.

[Solution note.

1.9. Whenever a polynomial $f(x)$ of degree n is tabulated for equal intervals of the variable x, the nth differences will be zero. If a table of values is given which consists of polynomial values that have been rounded off, then the differences will become close to zero and oscillate in sign. This point is discussed further in 4.10.

The following figures are derived from a cubic polynomial and have been rounded to four significant figures. Form a difference table and observe how the 3rd differences behave.

26·00, 29·64, 33·64, 38·03, 42·83, 48·05, 53·71, 59·83, 66·44, 73·54, 81·17

It is usual when constructing difference tables to omit the decimal points and leave them to be understood. Thus the first entry in the first difference in the exercise would be 364.

MULTIPLICATION

Learn how
  (i) to multiply a given number, e.g. **37·48**, by an integer such as **297**;
  operating first with the first figure (2, here).
  (ii) how to do the same more shortly as **300 − 3**, checking the process
  by the figures in the multiplier register.

2.1. Multiply (i) **57·88** by **967**
        and (ii) **9·67** by **5788**

Do each (*a*) by giving the full number of turns indicated by the
multiplying figures (e.g. 9, 6, 7) and (*b*) by using backward turns to
reduce the total.

Note that this calculation can be done in as few as seven turns.

Consider how you would have put the decimal point markers if the
calculation (i) had been **578·8** by **96·7**. Markers should appear (*a*) in the
setting register, where the figures 5788 are set up; (*b*) in the multiplier
register, where 967 appears *after* the calculation is done; and (*c*) on the
accumulator for the result.

Notice that the *decimal point markers do not in any way affect the
operation of the machine*, but only the interpretation of input and output.

For 2.2. If your machine allows it, learn how to transfer a result from
the accumulator back to the setting register.

2.2. Multiply **4·83** by **5·54** and the result by **22·3**. Check by carrying out
the multiplication in a different order. (If you have a back-transfer from
the accumulator, use it to set your first product, ready to form the triple
product.)

2.3. Evaluate **3·79 × 5·4 + 12·6**, first in this order; and secondly by
adding 12·6 (in correct position) into the accumulator first.

2.4. Evaluate **11 × 1·3 + 35 × 1·475 + 24 × 2·67** *first* with the integers as
multipliers, noting the final figure shown in the multiplier register; and
second with the decimal numbers as multipliers.

Notice that although the latter method takes longer it gives a result which might often be required viz.: the sum of the decimal numbers, as in the following example:

Determine the total cost and weight of a freight consignment of **1·3 tonnes at \$11 per tonne, 1·475 tonnes at \$35 per tonne, and 2·67 tonnes at \$24 per tonne.**

2.5. The heights of **100** boys are measured to the nearest inch and are found to fall into these intervals:

Inches	50–52	53–55	56–58	59–61	62–64	65–67	68–70
No. of boys	3	12	19	30	21	11	4

Ascribe to each boy, for convenience, the mid-height of the interval in which he comes; and work out the estimated total height $51 \times 3 + 54 \times 12$ and so on.

Hence find an estimate of the average height.

(Notice the check the multiplier register gives, that the numbers of boys have been correctly included in the calculation.)

The result is checked by the calculation of Ex. 2.6.

2.6. The data are as in Ex. 2.5, but the width of the interval $(3'')$ is taken as a unit of height throughout the work, and the heights are measured not from the floor but from $60''$ level as a standard; heights above being counted $+$ve and below $-$ve, thus:

Height in units	-3	-2	-1	0	$+1$	$+2$	$+3$
No. of boys	3	12	19	30	21	11	4

Evaluate the total of heights counted in this way, and the average.

Total is $3 \times (-3) + 12 \times (-2) +$ etc.

(Notice that since we cannot set negative numbers but we *can* multiply by them, the multiplier register reading will not this time give the number of boys. Will its value help us in any way to check that we have not omitted data or included any twice by mistake?)

Remembering how to interpret the final average in terms of units of $3''$, above the $60''$ level, verify that your result checks with that of Ex. 2.5.

2.7.   (*a*)   (*b*)

Evaluate $2·431 \times 4·7 - 0·09 \times 10·8$

    (i) by forming (*a*) in the accumulator first (but using either factor as multiplier according to convenience).

        and then forming (*b*) negatively;

    (ii) by forming (*b*) negatively first and then adding (*a*).

2.8. (*a*) Evaluate **3·9** × **4·88** and **6·2** × **4·88** in a single operation — by setting the first numbers far enough apart on the setting levers.

(Check: Verify that the sum of the results is 10·1 × 4·88.)

If your machine has a locking device for part of the accumulator, .learn how to use it.

(*b*) Repeat Ex. 2.8 (*a*), getting one result where you can lock it. Do so; then set the other, shift so that the two are in alignment, and accumulate.

If your machine has back-transfer from the accumulator to the setting-register, proceed as follows:

(*c*) Repeat Ex. 2.8 (*b*), but transfer back the one result instead of re-setting. (The other result, being locked, will not transfer.)

2.9. A number of parallel forces (or weights) W_1, W_2, etc. act at points on line Ox, drawn perpendicular to them, at points distant x_1, x_2 etc. from O. The sum (M) of the moments about O is $\sum (Wx)$, i.e.

$$W_1x_1 + W_2x_2 + \text{etc.}$$

and the sum (P) of the forces is $\sum (W)$, i.e.

$$W_1 + W_2 + \text{etc.}$$

Find in a single series of operations the values of M and P and calculate $\bar{x} = M/P$.

Value of W	4	3	5	2	6	(kg.)
Value of x	8·9	9·4	9·7	10·4	10·8	(metres)

Repeat the calculation, where x is measured from the 10 metre point, so that the values of x read:

$$-1·1, \ -0·6, \ -0·3, \ +0·4, \ +0·8$$

Verify that the position of the centre calculated in this way is the same as before.

ROUNDING. INTEREST. UNIT CONVERSION. ERRORS

This chapter is an extension and continuation of the previous one. What is new is principally arithmetical rather than mechanical.

We shall use the following abbreviations:

> **to 2D,** for 'to two decimal places',
> **to 4S,** for 'to four significant figures'.

The value **6·128,** expressed to 2D or 3S, is written as **6·13.** It is said to be *rounded* to 2D or to 3S.

The example given, of **6·128 → 6·13,** was rounded *up.* If **6·539** is rounded to 1D (or 2S) it is replaced by **6·5,** i.e. rounded *down.*

When a solitary final 5 is to be rounded, e.g. in **27·65,** we have to choose whether to round up or down. The most satisfactory rule is always to *round to an even number or zero,* e.g.

> **27·65** to 1D is **27·6**
> **48·95** to 1D is **49·0**

This procedure avoids systematic over- or under-estimates, since in the long run we shall round up as often as we round down.

Beginners may fall into either of two opposite types of mistake in relation to rounding; *either*

(1) to round values during the course of a calculation, thereby piling up errors in the final answer; *or*

(2) never to round at all, giving a final result to an unpractical number of figures.

1 The first course is always wrong unless one is working deliberately for an approximate answer, e.g. with a slide-rule.

2 No harm is done by carrying many figures, providing the final result is *either* rounded off *or, better,* is issued with calculated limits of error. We return to this in section **3.7.**

In practice most workers use two *guard-figures,* i.e. carry throughout the work two more digits than will be required at the end.

Rounding with money

A rather simpler rule applies when rounding sums of money, viz:

bear in mind the smallest coin which is used in payment: e.g. $4\frac{1}{4}$% of £100 is **£4·25,** but $4\frac{1}{4}$% of £1 is **£0·04** or 4p; and for £101 it is **£4·29.** This is by definition an *exact* answer. (In some situations a result will always be rounded down, e.g. $5\frac{3}{4}$% interest on £1 would be **5p.**)

3.1. (*a*) Round the following results to the number of places or figures indicated:

> (i) **37·285** to 2D, and to 3S;
> (ii) **4·2095** to 3D, and to 3S.

(*b*) Set the number **173·24** and multiply it by **7·9.** *Do not clear.* Note the result as displayed.

Without clearing, form the product of 173·24 and 8·2 (i.e. by making 3 extra turns with the carriage in a suitable position). Note as before. The time-saving machine process which you have just carried out is known as *continuation.*

(*c*) It is required to find **14%** of **381·78** and also to find the total after **381·78** has been increased by **14%.**

Set 381·78 and multiply by 0·14. Note the result. *Do not clear.* Add 381·78. (Since addition is recorded by a unit in the cycle register, this will now read as 1·14.) Note the second result.

Note. The fact that addition corresponds to multiplication by 1·0 is a valuable check on correct location. (The writer also uses the test $1·0 \times 1·0 = 1·0$ as a three-way check.)

3.2. A sum, deposited in a certain bank, yields **4%** interest after one year, and this if not drawn out it is added to the original sum. Interest at the end of a further year will be counted on the total sum credited to the customer at the beginning of that year, and so on (compound interest).

The sum of **$1247** is deposited. List the interest credited at the end of the year, and the amount; and so on (no money being drawn out) for a second and for further years up to ten. Set out as shown below and check by adding the total of interest payments to the original sum (or in detail, as for a difference table). This is the layout:

Date	Amount $	Interest $
Commencing date	1247·00	—
End of 1st year	1296·88	49·88

Notice that the result of the first stage of the calculation is the starting point of the next. If you are using transfer from accumulator to setting register, make sure to have the decimal point in the correct position before transfer.

3.3. A journey consists of a series of stages, each given in nautical miles. It is required to set out, in the form shown by the example below, the separate stage-distances and also the running total (both in *kilometres*). (The conversion factor is 1 nautical mile = 1·85 km.)

Distance		Running total
in n. miles	in km	in km
20·0	37·000	37·000
30·0	55·500	92·500
11·0	20·350	112·850

Note. We give a routine 3-figure decimal for km, not implying *accuracy* to 1 metre.

Data for your example are **20·8, 33·4, 17·4, 9·3, 18·5 n. miles.**

Method 1. Set the conversion factor on the levers. Obtain in turn each result for column 2. Subsequently re-enter these to form column 3. Check the final total by converting the total of data, e.g. 61 n. miles → 112·850 km. This checks intermediate results by implication.

Method 2. If your machine has an extra store you can use it to form the running totals concurrently. A simple form of store is a lock on one half of the product-register: you form the product in both halves, and only one of these is cleared.

3.4. (An extension of **3.3.**)

Find the greatest possible error in the final total of **3.4** due to each of the following causes:

    (i) because 1 nautical mile has been taken as 1·85 km instead of the international standard 1·852;

    (ii) because the data are only given to the nearest tenth of a nautical mile.   [Solution note.

3.5. Conversion of quantities expressed in non-decimal systems of units.

This may be carried out in either of two ways:

    (i) by expressing the data first in terms of one unit only (usually the smallest), e.g.

        8 stone 7½ lb = 8 × 14 + 7·5 lb = 119·5 lb = 119·5 × 0·454 kg

    (ii) by carrying out the conversion in two parts, e.g. to convert 7 lb 3 oz to kg rounded to 3D (i.e. to the nearest gramme).

We have   1 lb = 453·6 g and 1 oz = $\frac{1}{16}$ lb = 28·4 g.
Hence   7 lb = 3175·2 g
           3 oz =   85·2 g
           ─────────────────
           3260·4 g, giving 3·260 kg.

Make the following conversions:

(a) **5 ft 7$\frac{3}{4}$ in** to mm (1 ft = 304·8 mm). Give the result also in metres to 3D. (It is metric practice, except for areas, to use 3D; thus giving prominence to units by steps of 1000, e.g. km, m, mm and tonne, kg, g.)

(b) **57 gallons 3 pints** to litres, to 4S.
    (1 gallon = 4·546 litres and 1 pint = $\frac{1}{8}$ gallon = 0·5682 litres.)

(c) **3 miles 572 yards** to metres and state in km to 3D.
    (1 mile = 1760 yards and 1 yard = 0·9144 m.)   [Solution note.

3.6. Conversion of angles to and from degrees and minutes.

Although decimalisation of other units is now standard practice we continue with $\frac{1}{60}$ for units of time and in many instances for angles also. An important principle is, however, generally accepted: never to use more than *one subsidiary unit*, e.g.

(i) a time may be expressed as 2 hours 23·7 minutes (not using seconds). Similarly we might say 17 min 3·6 sec, though 17·06 min would be preferable:

(ii) an angle may be written as 43° 12′ or as 43·20°. The latter is preferable in all cases in which the angle might need to be converted to radians (or obtained from radian measure).

Carry out the following conversions:

(a) **3 h 57·2 min** to minutes and to hours, both to 4S;
(b) **57·296°** to degrees and minutes to nearest $\frac{1}{10}$′. (This is the magnitude of one radian, to 5S.)
(c) Convert **36° 52·1′** to degrees to 2D, and to radians to 4D.
(d) A ship moves **5 degrees 13·1 minutes** of arc along a meridian. Calculate the distance moved in nautical miles and in km. (1 minute of arc = 1 n. mile = 1·852 km.)   [Solution note.

For 3.7. To determine the greatest possible error in certain functions when the arguments are subject to errors up to a given magnitude. *Argument* is a better name than *variable* in this context.

(i) If a function of x is tabulated, the result required can be read off; e.g. if $x = 30·1°$, rounded to 3S, (i.e. between 30·1 ± 0·05) then the difference for 3 minutes gives the range of sin x to be

$$0.5000 \pm 0.0008$$

and this states the *acceptance* of the value 0.5000.

(ii) For simple cases the result can be written down.
If $x = 4.3 \pm 0.1$ and $y = 9.7 \pm 0.2$,
then

$$x - y = -5.4 \pm 0.3$$

(iii) For other cases an algebraic treatment can be used; e.g. for the product (xy) suppose the true values to be $x + \Delta x$, $y + \Delta y$; then neglecting the term $\Delta x . \Delta y$ the error is seen to be $x . \Delta y + y . \Delta x$.

(iv) Failing any other method it is possible to vary the values of the data, through their possible ranges, and to see directly what is the effect on the result. This is in fact what we have done in (i).

NOTE that in all practical contexts a numerical result, though fully checked, is of little value unless its acceptance is known.

3.7. Given $\mathbf{x = 4.52}$, $\mathbf{y = 0.73}$ (rounded values) calculate the values and acceptances of

     (i) $\tan y$, (y in radians);   (iii) xy
    (ii) $x + y$;   (iv) $x^2 y$

Check (iii), (iv) by direct calculation of an extreme value.

[Solution note.

A note on Flow Diagrams

In many instances we find that a specified calculating procedure is carried out again and again, each time with different data, until some prearranged condition is satisfied.

A clear way of setting out such a calculating programme is by a *flow diagram*. We show such a diagram below, devised for the solution to Ex. 2.5 on p. 6.

Instruction boxes are rectangular. *Decision boxes*, each containing a yes-or-no question are diamonds or, as here, rounded.

In this example we use abbreviations:

A for accumulator, C for cycle register.

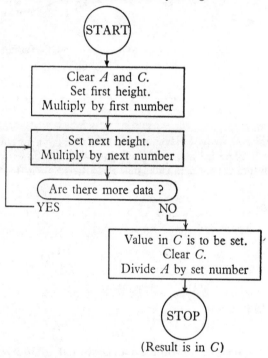

(Result is in C)

In later examples of flow diagrams the machine-details may be omitted, instructions being given in arithmetical or algebraic form.

FROM THIS POINT, ALL PRACTICE EXERCISES WILL BE FOUND ON RIGHT-HAND PAGES. NOTES FOR EXERCISES ARE ON LEFT-HAND PAGES

For 4.2. (Example is on right-hand page.) If you have a lock, do the example thus: set the integer in two positions before squaring; then one of the squares is cleared and the other is accumulated.

For 4.4.

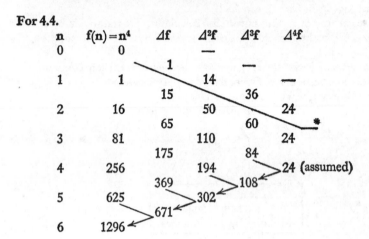

n	f(n)=n⁴	Δf	$\Delta^2 f$	$\Delta^3 f$	$\Delta^4 f$
0	0		—		
		1		—	
1	1		14		—
		15		36	
2	16		50		24
		65		60	
3	81		110		24
		175		84	
4	256		194		24 (assumed)
		369		108	
5	625		302		
		671			
6	1296				

The arrows show how, when a constant value for a certain order of differences is regarded as established, the table can be extended: but periodically a direct check must be made, e.g. here first at 10^4.

For 4.4. (*a*) The coefficients in these equations are those of the **Binomial Expansion**

$$(1+z)^n = 1 + \binom{n}{1}z + \binom{n}{2}z^2 + \ldots + \binom{n}{r}z^r + \ldots$$

For integral exponent *n*, they can be built up for successive values of *n* by **Pascal's Triangle**

1	1			
1	2	1		
1	3	3	1	

where each figure is the sum of the ones above and 'to the North-West', giving next:

1	4	6	4	1

(*b*) For non-integral exponent *p*, the coefficients must either be calculated as

$$\frac{p}{1} \qquad \frac{p(p-1)}{2!} \qquad \frac{p(p-1)(p-2)}{3!}$$

etc. or obtained from tables of interpolation coefficients.

4

SEQUENCES: EVALUATION. SUMMATION
AND DIFFERENCES FOR POLYNOMIALS

4.1. A sequence of numbers S_0, S_1, S_2,... can be evaluated in turn, given the value of S_0 and a rule connecting successive values. In this case the rule is $\mathbf{S}_n = \mathbf{S}_{n-1}(\mathbf{0 \cdot 8}) + 1$ and $\mathbf{S}_0 = \mathbf{0}$. The rule is called a *recurrence* formula.

Find the value of S_5 to 5D.

Verify by finding an algebraic formula for S_n (or failing that, for S_5) in as concise a form as possible.

(It may be convenient to write r for $0 \cdot 8$ and to substitute its value at the end.) Hence check your result. [Solution note.

4.2. Sum the squares of the first ten natural numbers (and record the individual values if it is possible to do this in one operation: see note). *Do not clear the multiplier register* between successive squarings.

What is the number accumulated in this register?

Verify this value and the sum of squares by using the algebraic formulae for

$$\sum_{r=1}^{n}(r) \quad \text{and} \quad \sum_{r=1}^{n}(r^2) \quad \text{for } n = 10.$$

4.3. Find by direct accumulation the value of

$$30^2 - 29^2 + 28^2 - 27^2 + \ldots + 22^2 - 21^2$$

Verify that the value is $10(30 + 21)/2$. Why? [Solution note.

4.4. Set out in a column as shown the values of $f(n) = n^4$ for $n = 0(1)10$ and differences. The top differences are regarded as linked to the first value of n, as shown,* and are written as $\Delta f(0)$, $\Delta^2 f(0)$ etc., e.g. $\Delta^3 f(0) = 36$.

(a) Verify that $f(2) = f(0) + 2\Delta f(0) + \Delta^2 f(0)$
$$f(3) = f(0) + 3\Delta f(0) + 3\Delta^2 f(0) + \Delta^3 f(0)$$
which are examples of the **Gregory-Newton formula**

$$\mathbf{f(n) = f(0) + \binom{n}{1}\Delta f(0) + \ldots + \binom{n}{r}\Delta^r f(0) + \ldots}$$

Clearly this would not be used to find $f(n)$ for integral values, but it is used for $f(p)$ where p is not integral.

(b) Use the formula $f(p) = f(0) + \binom{p}{1}\Delta f(0) + \ldots$, to evaluate $f(\frac{1}{2})$.

First list the values of $\binom{p}{1}$ to $\binom{p}{4}$.

Check that the result is the fourth power of $\frac{1}{2}$.

For 4.6 (a) $n^{[2]}$ represents $n(n+1)$; $n^{[3]} = n(n+1)(n+2)$ and so on, while $n^{[1]} = n$. (This example is for the more advanced student.)

(b) We can use the Gregory-Newton formula, with n integral, for *summation* of a polynomial sequence.

If we write $\sum_{r=0}^{k-1} f(r)$ as $S(k)$, then $f(k) = \Delta S(k)$.

We regard the f column as the first difference column of the function S.

Then $S(10) = S(0) + \binom{10}{1} f(0) + \binom{10}{2} \Delta f(0)$ and so on, continuing the

terms so long as the differences are non-zero. The term $S(0)$, as the sum of zero terms of the f sequence, is zero.

For 4.7. The *present value* of a payment P a year hence is

$$P/1{\cdot}04 = (0{\cdot}9615)P.$$

(Treat $0{\cdot}9615$ here as if it were exact.)

The present value of P to be paid n years hence is $(0{\cdot}9615)^n \, P$.

For 4.9. If a table of values of a quartic polynomial were required e.g. for $x = 0(0{\cdot}1)1$, it would only be necessary to calculate until the 4th difference is established, (see Ex. 4.4.) It is $(4!)a_0$ for unit interval, and more generally $(4!)a_0 h^4$ for interval h. The h^4 factor could be verified by rewriting the table of Ex. 4.4 for $0{\cdot}1$, $0{\cdot}2$ etc., which gives the same first column figures multiplied by 10^{-4}. In practice the *difference* figures would be written *without* decimal points, the last figure in every entry being understood to be in the 4th decimal place.

4.5. Sum the cubes from 11^3 to 20^3. Check by using the formula $\sum_{1}^{n}(r^3) = n^2(n+1)^2/4$ for $n = 20$ and for $n = 10$.

4.6. (a) Establish in general (or for $r = 2, 3, 4$) that if $f_r(n)$ is $n^{[r]}/r!$ then $\Delta f_r(n) = f_{r-1}(n+1)$.

Use this result to build up in turn the values of $f_2(n)$, $f_3(n)$ and $f_4(n)$ from $n = 0$ to 10.

Check directly for $n = 10$.   [Solution note.

(b) Verify the equation of note **4.6** (b) for $\sum_{n=1}^{10} f_4(n)$.

4.7. Find, to the nearest cent, the present value, reckoning interest at **4% compound**, of a series of **5 annual** payments of **$ 550**, the first payment being a year hence. Before deciding your method, refer to Ex. 4.1.

Check by formula $\dfrac{P(R^5-1)}{R^5(R-1)}$ i.e. $25P\{1 - (0.9615)^5\}$.

4.8. Evaluate to 3D $f(x) = a_0x^4 + a_1x^3 + a_2x^2 + a_3x + a_4$ where $a_0 = 1.00$, $a_1 = 0.87$, $a_2 = -0.61$, $a_3 = 1.20$, $a_4 = 2.00$ and $x = 1.25$.

Method I.

List the powers to 5D. (When x^2 has been found, leave it set to find x^3, x^4.) Evaluate term by term.

Method II.

Find $u_1 = a_0x + a_1$, then $u_2 = u_1x + a_2$ and so on. Finally u_4 is the required result.

Find the greatest possible errors in the result in cases (i) x supposed exact and the a's 'rounded', e.g. 1.00 means some value between 0.995 and 1.005; and (ii) the a's are exact, and x subject to an error ± 0.01. (See Ex. 3.7.)   [Solution note.

4.9. Construct a table of $f(x) = 2.50x^4 + 10.7x^2 - 0.89$ for $x = 0(0.1)1$ using *method II* for $x = 0(0.1)0.5$.

Build up thereafter by differences; and check $f(1)$ directly.

For 4.10. When the values of a polynomial function have been rounded off, or when the function is not precisely polynomial but approximates to a polynomial function over a certain range, then the differences may not become precisely constant or zero (note that $\Delta^5 f$ is zero in **4.4**).

The table below gives values of a function $f(x)$ rounded to 6S for $x = 0.50(0.01)0.58$.

x	f(x)	Δf	$\Delta^2 f$	$\Delta^3 f$	$\Delta^4 f$
0·50	2·00000				
		−3922			
0·51	1·96078		152		
		−3770		−11	
0·52	1·92308		141		5
		−3629		−6	
0·53	1·88679		135		−4
		−3494		−10	
0·54	1·85185		125		5
		−3367		−5	
0·55	1·81818		120		0
		−3247		−5	
0·56	1·78571		115		−3
		−3132		−8	
0·57	1·75439		107		
		−3025			
0·58	1·72414				

An error of up to 0·000005 in $f(x)$ will mean a maximum possible error of ± 1 in the column Δf, ± 2 in $\Delta^2 f$, ± 4 in $\Delta^3 f$ and so on, giving $\pm 2^{n-1}$ in $\Delta^n f$. In the table above the 4th differences are all within ± 8 of zero, suggesting that the function might be a cubic polynomial. In fact, $f(x) = \frac{1}{x}$, so it is possible to approximate to $\frac{1}{x}$ to 6S over this range using a cubic polynomial.

4.10. The following are the values for the function $f(x)$ where $f(x) = 2{\cdot}00x^3 - 6{\cdot}60x^2 + 9{\cdot}00x - 1{\cdot}00$ for $x = 1{\cdot}00(0{\cdot}1)2{\cdot}0$.

3·400, 3·576, 3·752, 3·940, 4·152, 4·400, 4·696, 5·052, 5·480, 5·992, 6·600

Construct a difference table for these figures as they stand. Then correct each of the given values to 3S and construct another difference table for the corrected values and note the oscillation of sign in the column $\Delta^4 f$. Check that the maximum deviation of $\Delta^4 f$ from zero is permissible.

For 4.12. Detection of errors by differencing.

If a table of values is known or believed to be those of a polynomial (or a function which can be so represented in that range) errors can often be detected by forming a difference table.

An error $(+e)$ in a single value will affect the differences according to a fan pattern thus:

Error in f	in $\varDelta$**f**	in $\varDelta^2$**f**	in $\varDelta^3$**f**
			$+e$
		$+e$	
	$+e$		$-3e$
$+e$		$-2e$	
	$-e$		$+3e$
		$+e$	
			$-e$

$\Big\}$ etc.

As an example we difference the following which are given as values of a function at equal intervals of the variable.

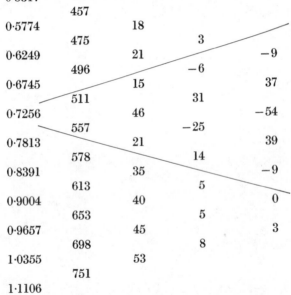

```
0·5317
          457
0·5774              18
          475               3
0·6249              21              -9
          496              -6
0·6745              15              37
          511               31
0·7256              46              -54
          557              -25
0·7813              21              39
          578               14
0·8391              35              -9
          613                5
0·9004              40               0
          653                5
0·9657              45               3
          698                8
1·0355              53
          751
1·1106
```

In this example, taken from the function $\tan x$, it looks as though the fourth differences should be close to zero. The characteristic fan pattern suggests that the value 7256 is wrong, and the numbers -9, 37, -54, 39, -9, being approximately in the ratio 1, -4, 6, -4, 1, suggest that the value is in error by being 9 too small; the correction may be effected by interchanging the last two digits.

4.11. The values of cosh x for $x=0\cdot0(0\cdot2)2\cdot0$, given to 5S are:

1·0000, 1·0201, 1·0811, 1·1855, 1·3374, 1·5431, 1·8107, 2·1509, 2·5775, 3·1075, 3·7622.

Show that the fifth differences oscillate and are within the range $\pm 2^4$.

4.12. The following are given as values of a polynomial function at equal intervals of the variable:

0·01, 0·01, 0·13, 0·73, 2·41, 6·10, 12·61, 23·53, 40·33, 64·81, 99·01, 145·21.

One value is in error; discover and correct it.   [Solution note.

4.13. The following are given as values for e^x for $x=0\cdot0(0\cdot1)1\cdot0$:

1·000, 1·105, 1·221, 1·350, 1·492, 1·649, 1·822, 2·041, 2·226, 2·460, 2·718.

Construct a difference table to find out which one of these values is incorrect; show also the corrected difference table.

Learn how to divide by 'teardown division', which is the long division method of paper arithmetic: good examples are

(i) 1·0000 by 7. What is the result to 4D?

(ii) 12·341 by 0·25 exactly.

Note particularly the locations of the data and the answers — quotient and remainder — in this process.

(*a*) The *dividend* (**N**) is required to be in the accumulator *at the left-hand end of it*. It is set, and transferred to the accumulator by a forward turn: the figure 1 which the turn causes to be shown in the multiplier register *must be cleared immediately*, and **N** itself is cleared from the set.

(*b*) The carriage is now shifted fully to the right, and the divisor (**D**) is set in the correct position for subtracting from the highest figures of **N**.

[The object of (*a*) and (*b*) is to ensure that the first quotient figure will appear at the left-hand end of the cycle register, and thus enable its full capacity to be used.]

(*c*) Division can now begin.

The whole process is shown in a flow diagram:

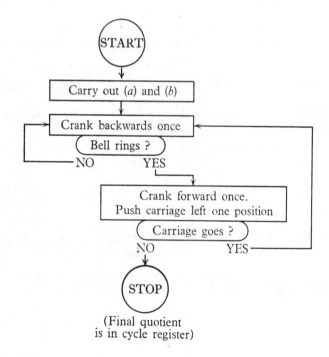

(Final quotient is in cycle register)

5

DIVISION. SUMMATION OF POWER SERIES

5.1. (*a*) Divide the answer to Ex. 2.2 ($p \times q \times r$) by each of its 3 factors: there will be no remainders. In each case check the decimal point of your answer by inspection of the product of the other two factors, i.e. test $(pqr)/r$ by inspection of pq.

(*b*) Find the reciprocals, to 4S, of **1·249, 0·08334, 750·2.** Check the figures of your answers by reciprocal tables, and then multiply each number by its 4S reciprocal.

Can you see why the discrepancies differ in magnitude?

[Solution note.

5.2. Calculate $\dfrac{2\cdot47 \times 39\cdot2}{43\cdot12}$

(*a*) in order, as you would read it; and

(*b*) by a method suitable if a series of numbers were to be reduced in the same ratio 39·2/43·12.

What is the greatest possible error in your result due to rounding errors in the data 2·47 and 39·2?

(It can be seen that the greatest possible error due to 43·12 is smaller, and you are invited to neglect it here.) [Solution note.

5.3. (i) An alloy is made from **10·3 kg.** of copper, **27·4 kg.** of nickel and **0·5 kg.** of tin. What is the percentage of each metal in the alloy?

Check that the separate figures total to 100, to the accuracy which is to be expected.

(ii) With the data of Ex. 3.7 find the acceptance of $z = x/y$. Verify that it agrees with the formula

$$(\varDelta z)/z = (\varDelta x)/x + (\varDelta y)/y, \quad \text{all numerators positive.}$$

Show that the same formula applies to Ex. 3.7 (iii). [Solution note.

For 5.4. Learn how to do 'build up' division, taking as examples again

(i) 1·000 by 7·0
(ii) 12·341 by 0·25

This method is preferred when the 'turns' can be used at the same time to multiply another number, e.g. in (ii) the number 2·43 could also be set (elsewhere on keyboard) and would give a result in the accumulator equal to:

$$2\cdot43 \times \frac{12\cdot341}{0\cdot25}$$

For 5.5 (i) If build-up division is used, notice carefully in which register the quotient is shown. In this problem the quotients are being accumulated, and this register must not be cleared after the first division.

(ii) Notice that teardown division can be used cumulatively if you look after the 1 which appears in the multiple register when a dividend is set in the accumulator. This 1 cannot be cleared *after* the second numerator is set.

It could be removed by making a back-turn *before* the figure is set.

For 5.7. (i) A possible method is to tabulate n and $1/(n!)$ for $n = 1\,(1)\,15$, to as many figures as the machine allows.

(ii) Could you shorten the amount of paper-work by accumulating reciprocals as they are formed? See Ex. 5.5 (i).

A flow diagram for this work:

5.4. Repeat part of Ex. 5.3, i.e. find the percentage of copper by 'multiplying up' the total weight to make the weight of copper and at the same time 'multiply' by 100 to give the percentage.

(As stated, this setting of 100 is unnecessary since result figures are given in the multiplier register; but if the point has been followed you can repeat the procedure to get in one move the answer to the following problem:

In **4.39** kg. of the alloy, how much nickel would there be, by weight? (3S).

Check your result by using your calculated % of nickel from Ex. 5.3.

5.5. Evaluate (if possible without getting both quotients separately and adding)

$$\frac{1 \cdot 74}{0 \cdot 231} + \frac{2 \cdot 48}{1 \cdot 75}$$

(Consider which method of division to use)

Check by repeating, forming the second fraction first.

5.6. In one year a country exports **57,100** tons of a commodity for **£238,400**; and in the following year **51,500** tons for **£253,100**. Find the % increase or decrease in each case and write them to the correct number of significant figures. Find also the % increase in the price, and show that the results are consistent.

5.7. Devise a method of evaluating e correct to 5D (working to 7D) by means of the series

$$e = 1 + 1/(1!) + 1/(2!) + 1/(3!) + \ldots.$$

Record your method in detail, and check your answer with the known value of e.

It is also checked by Ex. 5.8.

A self-contained check for this exercise, if method (i) is used, is to evaluate also

$$e^{-1} = 1 - 1/(1!) + 1/(2!) - 1/(3!) \text{ etc.},$$

for which all the terms are known. Then check $e^{-1} \cdot e = 1$ to the expected accuracy.

For 5.8.

Since x is the reciprocal of an integer, your method of Ex. 5.7 could readily be modified; but you should consider also how to programme if $x = 0.51$ say.

(If you list powers, N.B. method I of Ex. 4.8; x^4 can be set to get values x^5 to x^8, but cross-check by e.g. $x^5 \times x^3 = x^8$.)

For 5.9. The result, $\log 10 = 2.3026$, could be checked by evaluating e to the power 2·3026. A method would be

    (i) to find e^2 directly;
    (ii) to find $e^{0.1}$ by expansion and cube it;
    (iii) and $e^{0.0026}$ by a (very short) expansion;
    (iv) multiply all together.

How many figures to take in each part, and how to assess the final figure ($= 10$ nearly) are matters of nice judgment.

For 5.11. Notice that owing to the nature of the number $x = \frac{1}{5}$ you will probably prefer to find powers by multiplication for the first expansion, and division for the second.

5.8. Modify your method of Ex. 5.7 to find $\sqrt{e}$ by the series

$$e^x = 1 + x + x^2/(2!) + x^3/(3!) + \ldots, \quad \text{with } x = \tfrac{1}{2}.$$

Get the result correct to 6D. By squaring this result, to as many places as your machine allows, check against the value of e.

(Record the values of all terms of the series for use in Ex. 5.10.)

5.9. Using the series (valid if $-1 < x < 1$)

$$\log_e\left(\frac{1+x}{1-x}\right) = 2 \{x + x^3/3 + x^5/5 + x^7/7 + \ldots\},$$

evaluate (i) $\log_e 2$ (ii) $\log_e (5/4)$ and hence find $\log_e 10$.
(Consider carefully whether to get the later odd powers by dividing by $1/x^2$ or multiplying by x^2.)

5.10. Using the values of the terms found in Ex. 5.8 determine the sine and cosine of an angle equal to $\tfrac{1}{2}$ radian from

$$\text{Sin } x = x - x^3/(3!) + x^5/(5!) - \text{etc.}$$
$$\text{Cos } x = 1 - x^2/(2!) + x^4/(4!) - \text{etc.} \quad \textbf{to 5D.}$$

Square and add, to check these results.
Also convert $\tfrac{1}{2}$ radian to degrees, using $\pi = 3 \cdot 14159$ and compare with table values of sine and cosine.

5.11. Evaluate $\tan^{-1}(1/5)$ and $\tan^{-1}(1/239)$ in radians to 8D, using the series, valid for $-1 < x \leqslant 1$,

$$\tan^{-1} x = x - x^3/3 + x^5/5 - \text{etc.}$$

Hence find π, using $\pi = 16 \tan^{-1}(1/5) - 4 \tan^{-1}(1/239)$.

Describe your method, and state with reasons how many figures in your result you claim to be accurate (or better still, the uncertainty in the last figures obtained from your calculation).

Great emphaisis has been laid, in all the foregoing work, on the checking of results. This will have taught the reader those ways in which he is most liable to error, and (in places where the check shows an inevitable discrepancy) the limitations inseparable from the results at the best of times; but it is clear that there is a heavy duplication of labour in, for example, the checking of an isolated result like that of Note 5.9.

Two principal ways in which this work can be avoided are:

(1) Values of a function $f(x)$, computed for regular intervals of the variable, e.g. sin x at 10° intervals, can often be made to check each other by examining the successive differences. Functions which can be represented by polynomials over the range of the table will be found to have a difference column which is constant as nearly as the rounding off of the values of $f(x)$ will allow.

The method of detecting a *single* error in a table has been set out in Note 4.12.

(2) For isolated values it may sometimes be possible to check by a quick process e.g. the root of an equation by substitution, but it may be hard to judge whether the inevitable discrepancy, the 'residual', is small enough — see later Note 8.6.

Wherever possible an isolated value is found by an *iterative process* in which an approximation is successively refined again and again until it emerges unchanged from the process — to the required degree of accuracy. The next chapter is devoted to these methods, which are self-checking.

Square root.

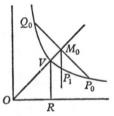

Graph of $y = N/x$. *See opposite page.*

6

ITERATIVE METHODS, INCLUDING STARTING PROCEDURE

Square root of a number N.

To find the square root of a number N to any required degree of accuracy, starting with a reasonable first approximation x_0. Calculate in turn the numbers $x_1 = \frac{1}{2}(x_0 + N/x_0)$; $x_2 = \frac{1}{2}(x_1 + N/x_1)$ and so on, the general relation being $x_{n+1} = \frac{1}{2}(x_n + N/x_n)$.

After a few iterations the successive values of x are found to differ by extremely small (and diminishing) amounts. When successive results agree to the degree of accuracy required, the answer has been obtained; e.g. if x_7 and x_8, rounded off to 5S, are both equal to 1·4142, then this is the required square root to 5S.

The method can be illustrated graphically by the sketch of the graph of $y = N/x$ on the opposite page.

The points P_0 $(x_0, N/x_0)$ and $Q_0(N/x_0, x_0)$ both lie on the graph and are images of each other in the line $x = y$, on which their mid-point M_0 lies. The required root $x = N/x$ is the x-coordinate of the point V, and is equal to OR in the figure.

It will be seen that the x-coordinate of M_0 is the mean of those of P_0 and Q_0, i.e. $\frac{1}{2}(x_0) + \frac{1}{2}(N/x_0)$, which is x_1. We therefore drop a perpendicular from M_0 to the x-axis, to cut the curve in $P_1(x_1, N/x_1)$ whose image is Q_1, and the midpoint M_1 of P_1Q_1 gives in the same way P_2; and so on.

It is clear from the figure that the successive perpendiculars tend towards the position VR, and the x-values to the required root.

Verify, on the same figure, that the values also converge to $\sqrt{N}$ if x_0 is *smaller* than the true value.

(This analysis shows the values x_1, etc., as steadily decreasing; but, as in all digital processes, rounding may affect this unpredictably.)

6.1 Use the method described to evaluate $\sqrt{10}$ to 4D starting with $x_0 = 3·0$. Note your programme of operations in detail. [Solution note.

Note for 6.2 Newton's Method for finding the root of an equation.

This is an iterative process of very wide application which enables us to obtain to any required degree of accuracy the value of a root of the equation $f(x) = 0$, given a sufficiently close approximation (x_0) to this root.

We write
$$x_1 = x_0 - \frac{f(x_0)}{f'(x_0)},$$

where $f'(x)$ is the first derivative of $f(x)$; and in general

$$x_{n+1} = x_n - \frac{f(x_n)}{f'(x_n)} . \text{ Newton's Formula}$$

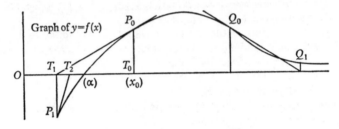

The point P_0 is $[x_0, f(x_0)]$ and the gradient at this point is
$$T_0 P_0 / T_1 T_0 = f'(x_0),$$
$$\therefore \qquad T_1 T_0 = T_0 P_0 / f'(x_0) = f(x_0) / f'(x_0)$$
$$\text{Thus} \qquad OT_1 = x_0 - \frac{f(x_0)}{f'(x_0)} = x_1$$

and a similar construction gives $OT_2 = x_2$ and so on.

Notice however that if we start at Q_0, beyond the turning point, the next result is worse.

It will be seen that the vanishing of $f'(x)$ at a point *near* P_0 will demand more accuracy in the starting value x_0. It is such considerations, as the graph above will make clear, which indicate what is involved in the words 'sufficiently close'.

N.B. It must not be supposed that Newton's formula fails for a multiple root: the difficulty occurs for simple roots which are nearly equal, or for a root with a turning point very near it.

Newton's method can be adapted to give a slightly different method for obtaining the root of the equation $f(x) = 0$. The adapted formula is less efficient in that it may require more iterations to reach the desired accuracy; on the other hand it is a simpler formula to use for computation purposes, a fact which will save time.

6.2 (a) Notice that the formula given by Newton may require to be simplified before it is actually used for calculation, e.g. to obtain a square root we may write $f(x) \equiv x^2 - N$, which gives

$$x_1 = x_0 - \frac{(x_0{}^2 - N)}{2x_0}$$

Simplified algebraically this gives the formula used in Ex. 6.1.

6.2 (b) Obtain an alternative formula for the square root, by writing $f(x) \equiv 1/x^2 - 1/N = 0$, in the form:

$$x_{n+1} = x_n \left(1 \cdot 5 - \frac{x_n{}^2}{2N} \right)$$

A simple machine programme would be as follows:
First calculate and note the reciprocal of $2N$: clear.

Iteration (i) form the product $x_n(1/2N) = P$;
  (ii) set P and clear accumulator (manually or by direct transfer if available;
  (iii) multiply P by $(-x_n)$: do not clear;
  (iv) add $1 \cdot 5$, to get value of bracket B;
  (v) set B and clear accumulator as in (ii);
  (vi) multiply B by x_n to give $x_{n+1} = X$: note this value;
  (vii) set X and clear accumulator.

Thereafter the process is repeated as often as is required.
Use this programme to solve Ex. 6.1.
Compare the two procedures for (*a*) convenience, and (*b*) rapidity of convergence.
It will be appreciated that a machine can be made to carry out the whole process unaided if it has the following facilities:

  (i) for *storing* numbers which are required to be noted in the method above;
  (ii) for *feeding* in or *transferring* numbers, as required for processing;
  (iii) for *testing* whether x_n and x_{n+1} are or are not equal within the required limits, and giving an appropriate *YES* or *NO* signal;
and (iv) a *control unit*, to ensure that the prearranged programme is carried out, and repeated so long as the signal is *NO*.

See also p. 82: Programming and the Computer.

6.2 (c) Use the adaptation of Newton's method to obtain yet another formula for the square root of N, namely

$$x_{n+1} = x_n - \frac{(x_n{}^2 - N)}{2x_0}.$$

In the formula

$$x_{n+1} = x_n - \frac{f(x_n)}{f'(x_n)}$$

we may assume that, provided the curve is not turning too quickly, the value of $f'(x_n)$ will not change by very much as a percentage of itself from iteration to iteration. We thus replace $f'(x_n)$ in the formula by $f'(x_0)$ and obtain

$$x_{n+1} = x_n - \frac{f(x_n)}{f'(x_0)}$$

The diagram illustrates the new method. $P_0 T_1$ is the tangent at P_0, and $P_1 T_2$, etc. are parallel to $P_0 T_1$. The distance $O T_n$ is given by x_n.

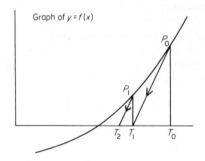

Graph of $y = f(x)$

In practice, the value of the reciprocal of $2x_0$ (which, in this particular case, is the value of $f'(x_0)$) is worked out and then used in all the iterations.

6.3. Derive from $f(x) = 1/x - N$ a formula and a programme for finding a reciprocal, and use it to find $1/31$ to 6S, starting with $x_0 = 0.03$. Notice that division is not used.

6.4. Starting with $x_0 =$ the value of the square root of 11 given by 4-figure tables, find x_1 and x_0, using either formula from page 31. State the best result obtainable from this work.

6.5. Starting with 3.16 as the first approximation to the square root of 10, compare the formula

$$x_{n+1} = x_n - \frac{x_n^2 - 10}{2x_n}$$

with its adaptation according to the method of **6.2.(c)**

$$x_{n+1} = x_n - \frac{x_n^2 - 10}{6.32}$$

as methods for finding the square root of 10 to 6S.

For 6.6.

Formulae are:

$$x_{n+1} = x_n\left(\frac{4}{3} - \frac{x_n{}^3}{3N}\right) \quad \text{and} \quad x_{n+1} = x_n\left(\frac{2}{3} + \frac{N}{3x_n{}^3}\right)$$

For 6.7.

Iteration by direct formula for equation x = F(x)

This works, writing $x_{n+1} = F(x_n)$ provided that the gradient $F'(x)$ in the region used, is numerically < 1. This can be seen by drawing figures as below for various cases. Sketch $y = x$ and $y = F(x)$.

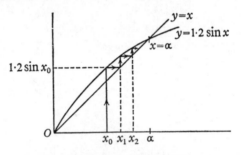

The graph shows how the successive approximations are related to one another. It is clear that this is not an accelerating convergence like Newton's, which frequently *doubles the number of significant figures* at each iteration. By contrast, this method gives steps which diminish in nearly constant ratio.

6.6. Using either the method of Ex. 6.1 or of Ex. 6.2 to get a Newton formula for a root, find to 4S the cube root of 28. $(x_0 = 3)$.

Without completing a second calculation, consider the advantages of the above method compared with the use of the Binomial expansion:

$$28^{1/3} = 27^{1/3}(28/27)^{1/3} = 3(1 + 1/27)^{1/3}$$

$$= 3\left\{1 + \frac{1}{3}\left(\frac{1}{27}\right) + \frac{1}{2!}\left(\frac{1}{3}\right)\left(-\frac{2}{3}\right)\left(\frac{1}{27}\right)^2 + \ldots\right\}$$

6.7. Find a positive root of the equation $x = 1 \cdot 2 \sin x$ (x in radians) to 4S by direct iteration, using the table of sin x for x in radians, and the formula

$$x_{n+1} = 1 \cdot 2 \sin x_n$$

If you have only a table of sin x for x in *degrees*, you could* use $x_{n+1} = \sin x_n$ to get a root in degrees, and convert to radians. (Notice that 4S is almost the same accuracy requirement in radians as in degrees. Why not *exactly* the same?)

*It will illustrate the method equally well to solve the (different) equation $x = 80 \sin x$ to 4S where x is in degrees; but you should have access to a book of tables which includes functions of a radian argument. See note on Tables, Appendix IV.

For 6.8. To find a first approximation to a root of f(x) = 0.

(I) Method of interpolation. (See footnote*)

By trial we find values X, $(X + \Delta X)$ such that $f(x)$ has opposite signs for the two values. In the figure $f(X)$ is $-P$ and $f(X + \Delta X)$ is Q.

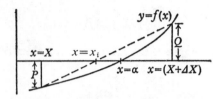

It will be seen that the chord intersects the x-axis at a better approximation (x_i) than either X or $X + \Delta X$.

The gradient of the chord $= \dfrac{P}{x_i - X} = \dfrac{Q + P}{\Delta X} = \dfrac{\Delta f}{\Delta X}$

$\therefore \qquad x_i - X = P(\Delta X / \Delta f) = -f(X)[\Delta X / \Delta f]$

$\therefore \qquad x_i = X - f(X)[\Delta X / \Delta f]$

a form which will suggest the programme of computation.

(**II**) Comparison of the diagrams for 6.2 and 6.8 shows that if x_i is too small, the value obtained by one Newton iteration from X or $X + \Delta X$ is too large; while a downward concavity reverses this situation. This suggests the rule:

    (a) From X and $X + \Delta X$ calculate the interpolate value x_i as above;

    (b) from the better of X and $X + \Delta X$ calculate x_j by one preliminary iteration;

and (c) take $x_0 = \frac{1}{2}(x_i + x_j)$ as the start for the main iteration.

---

* **Interpolation** is the calculation of an intermediate function value. This example is of *linear, inverse* interpolation. 'Linear' explains itself, and 'inverse' is clear if we consider the solution of say $\sin \theta - 0 \cdot 6 = 0$: we might estimate a value of $\theta = \sin^{-1}(\cdot 6)$ from $\sin^{-1}(\cdot 5)$ and $\sin^{-1}(\cdot 7071)$.

6.8. It will have been noticed that an iterative method is very good *once it has been well started*. A good start may save several iterations.

A carefully drawn graph gives a good start, but judiciously calculated function values in the right region can be used without graphing by the **method of interpolation** (see note on left-hand page).

Example: To find, to 4S, the real root between **3** and **4** of the equation $x^3 - 3x - 24 = 0$.

We could take $f(x) \equiv x^3 - 3x - 24$, for which $f(3) = -6$, $f(4) = 28$ but with steadier gradient is $g(x) \equiv x^2 - 3 - 24/x$.

(i) Interpolate between 3 and 4 by method (I) of left-hand page and use this as starting value for Newton, with either f or g. (ii) Repeat the solution using method (II) to find a starting value.

6.9. Find to **4D** a solution of $e^x = 3x$, given that:
$$e^{1\cdot5000} = 4\cdot4817$$
$$e^{1\cdot5100} = 4\cdot5267$$
$$e^{1\cdot5200} = 4\cdot5722$$

Method I. Let $x = 1\cdot51 + h$, and remember that $e^{(1\cdot51+h)} = e^{1\cdot51}e^h$

Method II. Work as in Ex. 6.2 (ii) for $f(x) \equiv e^x - 3x$.

Here as in Method I, values of e^x not given in the table must be found by expansion.

The check is in any case by expansion.

Notice that if the function f were known only by its table we should have to make assumptions about its behaviour which could only be tested by higher differences. See Ex. 4.4 (*b*); and for further reference: Interpolation and Allied Tables (H.M. Stationery Office).

Method III. See Ex. 6.7 and Note 6.7.

In the direct form $x_{n+1} = e^{x_n}/3$ this method fails to converge.

The form $x_{n+1} = \log_e(3x_n)$ can be used, starting with $x_0 = 1\cdot5089$ derived from the value $\log_e(4\cdot5267) = 1\cdot5100$.

The reversed equation is, however, better solved by Newton's method with
$$f(x) = x - \log_e(3x)$$

Use $\log_e(a + h) = \log_e a + h/a + \text{etc.}$

7

STATISTICAL AND ALLIED CALCULATIONS

Statistics is a branch of applied mathematics which is concerned with deriving from a set of numerical data — often a very large set — certain numerical measures which will indicate the principal features of the set, either considered alone or in comparison with another set.

It is helpful sometimes to remember that the word statistics can be considered as a plural. A 'statistic' is a figure worked out from the data, e.g. from a census. It might for example be the proportion of males in the population of the country; or simpler still, the total population itself.

A set of numerical data usually consists of the measurement of one characteristic of every individual member of the group: it could be for example the height (and possibly the weight as well) of every man in a certain group called up for military service.

(1) The first statistic is the *number* of individuals in the group (symbol N). This should always be recorded.

(2) The second statistic is the *mean value* of the measure which is being considered, e.g. the mean height of the men. If x_1 is the height of the first man, x_2 of the second, and so on to x_N for the last, then the mean value is

$$\bar{x} = \frac{x_1 + x_2 + \ldots + x_N}{N} = (1/N)\sum_{r=1}^{N} x_r \tag{1}$$

The formula is sometimes used in this form, but usually two modifications arise:

(*a*) The measurements are made to only limited precision (e.g. $\frac{1}{2}''$, in height) and for convenience we may wish to group results to an even wider interval. In either case a number of men belong to the same class (e.g. 23 have $x = $ '65', 28 have $x = $ '66', where 65 means a height of 65 or above but lower than 66). The number 23 would be called the *frequency* of the value measured as 65.

Then $$\bar{x} = \frac{f_1 x_1 + f_2 x_2 + \ldots + f_c x_c}{N} = (1/N)\sum_{r=1}^{c} (f_r x_r) \tag{2}$$

where x_1 is the value and f_1 the frequency for the first *class*, and c is the number of classes. This calculation is much shorter, but presupposes the classification of the data — a process which serves another important purpose, viz.: to put the data in a form (pictorial or otherwise) in which their general nature can be exhibited to the eye.

The shape of the data is often best shown by having about nine groups, e.g. for heights of men (i) under 5′ (ii) 5′ but under 5′2″ (iii) 5′2″ but under 5′4″ and so on to (ix) 6′2″ and over. The mean obtained by grouping in this way, with 10 terms in equation (2), would be slightly different from that given by using all the possible groups, but would be clearer to see and easier to work with.

(Another consideration affecting the mean is discussed on p. 40.)

7.1. The following are the heights to the nearest inch, of a series of individuals. Classify them (a) into inch classes; and (b) into classes 50–52 inclusive, 53–55, 56–58, etc.

Values of height:

55, 57, 60, 64, 52, 60, 68, 62, 57, 58, 55, 61, 62, 58, 54, 69, 68, 62, 50, 59, 66, 57, 61, 65, 54, 61, 60, 58, 54, 63, 70, 59, 54, 57, 53, 64, 56, 61, 63, 54, 64, 62, 58, 65, 57, 58, 61, 66, 66, 59, 61, 56, 60, 65, 64, 60, 67, 64, 62, 65, 59, 61, 59, 63, 62, 68, 58, 61, 53, 64, 59, 62, 50, 59, 67, 61, 61, 58, 59, 62, 56, 59, 64, 60, 65, 57, 58, 61, 60, 54, 53, 60, 55, 59, 63, 59, 62, 57, 64, 56.

It will be necessary to use a systematic method of classifying, e.g. to give a column to each class and to enter each number in turn as a stroke in its column (as well as crossing the number off the list). It is convenient to put each fifth stroke across the rest (thus, ⧻⧻) so that the pattern of strokes at the end indicates the count clearly and accurately. The stroke total should agree with a direct count of the data (but this is no safeguard against entry in a wrong column).

After classifying, calculate $\bar{x}$ (a) from the inch-classes, and (b) from the 3-inch classes. Notice the small value of the discrepancy. (In case (b) the value of x taken for the class is to be the mid-value of the class. See Ex. 2.5.)

Refer now to Ex. 2.6 for a simpler method of calculating by using a class-centre as origin and the class-interval as unit. Repeat this, taking values of x *from* **0** *to* **6** instead.

The calculated mean: dependence on the type of measurement.

It should be noted that the calculated mean has to be understood with reference to the treatment of the original measurements. If the measurements were taken *disregarding all fractions of an inch* the calculated figure is too low by an amount which can be taken as $\frac{1}{2}''$; but if they were *to the nearest inch* it can be accepted as it stands.

For 7.3.

The third statistic which we shall consider provides a measure of the 'spread' of the data. Similar electric lamps made by two makers P and Q might prove, by a test of two equal sample groups of 100 (collected in some random way), to have the same mean life; but if P's output contained appreciable numbers (i) with less than one third and (ii) with over twice the average life, then this pattern would be less satisfactory because less dependable.

It would be unreliable to measure the difference between the best and worst lamp. A possible method is to measure the difference between the 25th and the 75th in order of merit in each group; but the method preferred is to find the *standard deviation* (s) of each group.

We compute the deviation of each value x from the mean ($\bar{x}$); sum the squares of these deviations and divide by N, the total number of instances, to get the **variance V.**

$$V = (1/N)\sum(x - \bar{x})^2 \quad \text{summed over all instances;}$$

but since there are f instances of each value x, we in fact compute

$$V = (1/N)\sum\{f_r(x_r - \bar{x})^2\} \quad \text{summed over the number of classes.}$$

Then s is defined to be $\sqrt{V}$.

This is the true standard deviation.

It will be seen that the direct calculation involves all the separate deviations being measured from $\bar{x}$, a non-integral value. An indirect method will be seen later which avoids this complication.

7.2.

You are given the following accident figures for a year in a certain district:

$n =$	0	1	2	3	4	5	6	7	8	>8
No. of days in which n fatalities occur	30	99	115	81	28	9	3	0	1	0

Find (i) the total number of fatalities and (ii) the mean number of fatalities caused per day.

Notice that the original data were a series of dated records each stating the number of fatalities caused on the day in question. Classified they appear as above, the frequency f being the number of times the number n appears in the records. (n is used in place of x since it is of its nature integral, though this not true of $\bar{n} = (1/N)\sum_{r=0} (rf_r)$.)

The summation is carried far enough to cover all cases: as always $N = \sum f_r$ over all classes (i.e. including f_0).

7.3. Referring to Ex. 2.5 and Ex. 2.6 (which is a restatement in more convenient form) we will show how to arrange a calculation of s directly.

Here $\bar{x} = +0.03$

Set out the values of $x - \bar{x}$ for each class, with the frequency of the class; preferably in columns with the latter first:

f	$x - \bar{x}$	$f(x - \bar{x})$	$f(x - \bar{x})^2$	
3	-3.03	-9.09	27.54	
12	-2.03	-24.36	49.45	Complete the last
19	-1.03	-19.57		column: **sum it**
30	-0.03	-0.90		(S_M): **Calculate**
21	$+0.97$	$+20.37$		**V** $= S_M/N$
11	$+1.97$	$+21.67$		**and** $s = \sqrt{V}$
4	$+2.97$	$+11.88$		This is in the units

$\sum f = 100 = N$ Sum $= 0.00$ $=$ used for measuring x: convert to inches.
(check, not usually exact)

For 7.4.

Standard deviation using an origin other than the mean.

The same drill is carried out with values calculated from a more convenient origin at a distance b from the true mean. The sum S of the squares so obtained is *always larger* than from the mean. The value of s is obtained from the equation

$$s^2 = S/N - b^2$$

This result is exact, and is used in either of two ways, according to the chosen point for measuring the deviation of each value:

(A) using the *central* class-position as origin: this has the advantage that the correction b^2 is rather small — it may even be negligible to the accuracy required.

or (B) using an *end* class-position: this has the advantage that all the numbers in the calculation are positive.

Examples: (A) Ex. 7.3 gives, taking zero as the origin

f	x	fx	fx²	
3	−3	−9	27	($b = -0.03$)
12	−2	−24	48	
etc.				

Sums **N**=100		+3	195 = S_0
		= $N\bar{x}$	

Then $S_0/N = 195/100 = 1.95$ and $s^2 = 1.95 - 0.0009$.

(B) Using the end class as zero, as in the Ex. 7.2. (Write x for n)

f	x	fx	fx²	
30	0	0	0	As **x** is already measured from the end we need not change it.
99	1	99	99	
115	2	230	460	
81	3	243	729	
28	4	112	448	*False* $s^2 = $ **2133/N**
9	5	45	225	and $\bar{x} = $ **755/N = b**
3	6	18	108	Hence complete to **find**
0	7	0	0	**true** s^2
1	8	8	64	
366 = N		755	2133	

7.4.

Find the standard deviation of the following classified data. (They represent the weight distribution of 100 boys, grouped into 7 classes.) The value associated with each class is given the symbol y.

y	-3	-2	-1	0	$+1$	$+2$	$+3$
f	3	11	21	28	25	10	2

Obtain s by two methods,

 (i) with the origin as shown: and

(ii) moving the origin to the left-hand class so that y has values from 0 to 6: you should check on obtaining $\bar{y}$ in case (ii) that it is consistent with the position found for the mean in (i).

7.5. Check your value of s^2 for the accident figures on the left-hand page, by recalculating with x measured from an origin near its mean, so that x reads down the column from -2 to $+6$ i.e. method (A).

7.6. The superiority of method (B) for machine-work is shown by the following programme for the accident figures.

 (i) Set the first value of x in position 6 and x^2 in position 1. (Positions are counted from the right in the setting register.)

 (ii) Make a number of forward turns equal to the first value of f.

(iii) Clear the setting register *only*.
   f is now shown in the multiplier register, and in the accumulator fx and fx^2 separately.

(iv) Continue as above, with the next value of x and its square, and the corresponding f. These are therefore accumulating, and all the required sums are obtained together.

Do this with the figures in (B) opposite and verify the column totals.
Do the same with Ex. 7.3 taking $x = 0$ to 6.

For 7.7.

Product-moment correlation.

We have considered already the distribution of height and of weight, in suitable class-intervals, of a certain 100 boys.

It would have been possible from the data to exhibit the distribution in respect of both attributes at once, as shown in the grid, Ex. 7.7, where f (the figure in any location of the grid) shows the number of boys who have *both* the height *and* the weight classification corresponding to that location.

We should expect the boys who are above average in one respect to be so also, more often than not, in the other; i.e. we should expect + values of x and y to go often together, and also − with −. Thus the sum of (xy) products over all the individuals, if strongly positive, would indicate a positive correlation and could be used as a measure of this: we therefore calculate $\sum(fxy)$; and $(1/N)\sum(fxy)$ is called the **product-moment** about the origin i.e. about the centre of class (0, 0).

This calculated result needs to be corrected because the origin used is not at the true mean: the corrected value is

$$(1/N)\sum(fxy) - \bar{x}\bar{y} \quad \text{(to the true mean)}$$

(Note that the correction term $\bar{x}\bar{y}$ is not necessarily positive in this case.)

This result is of more value if it is divided by s_x and by s_y the standard deviations for height and weight (expressed still in terms of class-interval as a unit).

The final result is called the **product-moment correlation co-efficient** of the data. Calculated as described, it is bound to lie between + 1 (when all the results lie in the NW–SE diagonal locations) and − 1 (for the other diagonal).

7.7.

		HEIGHTS (x) in intervals of 3"							
		-3	-2	-1	0	$+1$	$+2$	$+3$	**Totals**
	-3	1	2						3
	-2	2	6	2	1				11
WEIGHTS	-1		3	12	3	2	1		21
(y)	0		1	4	17	3	3		28
in 5-lb	$+1$			1	7	11	4	2	25
intervals	$+2$				2	5	3		10
	$+3$							2	2
	Totals	3	12	19	30	21	11	4	100

To evaluate $\sum (\mathbf{fxy})$ proceed as follows:

(i) For each value of x, e.g. for $x = -3$, sum the values of (fy) corresponding to it i.e. in this case $(-3)(1) + (-2)(2) = -7$.
We have worked down the first column, and its contribution to our total is $(-3) \times (-7) = 21$.
Work similarly for each column and put the result at the foot in an extra row.
The sum of this row is the sum required.

(ii) For each value of y, e.g. for $y = -2$, sum the values of (fx) corresponding to it, i.e. here $(-3)2 + (-2)6 + (-1)2 + (0)1$.
The contribution of this *row* is $(-2) \times (-20) = 40$, which is put in an extra column to the right of the row.
The sum of this column should also be $\sum (fxy) = P$.

To determine $\mathbf{r} = (\mathbf{P/N} - \mathbf{\bar{x}\bar{y}})/s_x s_y$ use the values already found for $\mathbf{\bar{x}}$, $\mathbf{\bar{y}}$, s_x, s_y *in terms of class-intervals as units*, since the P calculation has been done in these.

Repeat the calculation with the same data but with the origin in the top left-hand square, so that all the values of x, y and of $\bar{x}$, $\bar{y}$ are increased by 3 and none are negative.

It is possible as a 'tour de force' to set (fx) separated from (fy), to multiply by $x000y$, and to accumulate thus at the same time (fx^2), (fxy) doubled, and (fy^2). Try this out with simple figures first.

For 7.8.

A new sidelight on standard deviation is obtained if the idea of moment of inertia is understood. Refer to Ex. 2.9 which is now to be regarded as a set of *masses* on a weightless bar.

The centre of mass, distance $\bar{x}$ from the origin, is given by

$$\bar{x} = \frac{\sum(mx)}{\sum(m)},$$

so that m takes the place of f in our statistical formula, and the total mass M takes the place of $\sum f = N$.

The calculation of moment of inertia, about an axis perpendicular to the bar through the origin, is the finding of $\sum(mx^2)$, and this when divided by M gives $k_0{}^2$, the square of the radius of gyration about O.

If we need the radius of gyration about a parallel axis through G, the centre of mass, we write

$$k_G{}^2 = k_0{}^2 - (OG)^2 = k_0{}^2 - \bar{x}^2 \quad \dots\dots\dots\dots\dots\dots\dots(1)$$

Thus there is an exact parallel with standard deviation, and the correction we have applied is like the application of the Theorem of Parallel Axes. The proof of this theorem for the special case of masses in a line would exactly correspond to the proof for s^2:

If $x = \bar{x} + z$ so that z is the deviation of the given value from the mean;

then $$fx^2 = f\bar{x}^2 + 2\bar{x}fz + fz^2$$

We now sum for all the values:

$$\sum(fx^2) = \bar{x}^2\sum f + 2\bar{x}\sum(fz) + \sum(fz^2)$$

Now $\sum f = N$; and $\sum(fz) = 0$ since the z's are measured from the mean. Furthermore the first and last terms of the equation are the sums of squares, S_0 and S_M from the origin and the mean respectively,

$$\therefore S_0 = N\bar{x}^2 + S_M = N\bar{x}^2 + Ns^2$$
$$\therefore s^2 = S_0/N - \bar{x}^2 \quad \dots\dots\dots\dots\dots\dots(2) \text{ Cf. (1)}$$

7.8.

Determine the radius of gyration of the masses in Ex. 2.9 (i) about the origin
  and (ii) about the centre of mass directly.
  Check by using the theorem of parallel axes.

7.9.

Further examples, such as the Poisson distribution, ideally suited for machine work will be found in statistics text-books.

An interesting exercise is the χ^2 **test,** to decide whether the characteristics of a given sample may be considered consistent with a certain assumed distribution of this characteristic in the 'population' from which the sample is drawn. An elementary example of this test is the following:

A die was shaken 120 times, and the distribution of scores was (1) 12, (2) 23, (3) 19, (4) 24, (5) 26, (6) 16.

On the assumption that the die is unbiassed, the expected value would be 20 for all six numbers.

Writing a_i for the actual and e_i for the expected frequency for the ith class, χ^2 is defined by the expression

$$\sum \frac{(a_i - e_i)^2}{e_i} \quad \text{summed over the classes.}$$

The larger the value of χ^2, the less likely is the result to have arisen by chance for a true die.* For this case

$$\chi^2 = \frac{8^2}{20} + \frac{3^2}{20} + \frac{1^2}{20} + \frac{4^2}{20} + \frac{6^2}{20} + \frac{4^2}{20} = 7 \cdot 10$$

The χ^2 tables show that in a large number of repetitions of this experiment one could expect χ^2 to be at least as large as this in about 21% of them, for a true die. By common practice the assumption (true die, in this case) is rejected only if the calculated χ^2 value is 'rarer' than 5%; and is in some doubt if it is rarer than 10%.

[* The six a-values are not all independently variable, since their sum is fixed: the appropriate line of the χ^2 table is that for '5 degrees of freedom'.]

A further trial with the same die gives 14, 19, 25, 23, 24, 15. Does this make it more likely that the die is biassed? [Solution note.

8

SOLUTION OF LINEAR EQUATIONS

In this chapter we can only give the reader, by means of simple examples free from pitfalls, an outline of the two main lines of approach, viz: elimination (Exx. 8.1 to 8.5) and successive approximation (Exx. 8.6 to 8.9).

The numerical process of solution involves only the *coefficients* and the lay-out consists of numbers only. We set out the skeleton data — the coefficients and the absolute term — together with an additional summation column. Thus, for the equations

$$5 \cdot 23x_1 + 0 \cdot 47x_2 = 2 \cdot 63$$
$$11 \cdot 25x_1 + 2 \cdot 39x_2 = 3 \cdot 08$$

we write

	x_1	x_2		S
(Equation E)	5·23	0·47	2·63	8·33
(Pivotal, P)	11·25	2·39	3·08	16·72
(Equation E')	(−0·0001)	−0·6411	1·198	0·597

$$\therefore \quad x_2 = -1 \cdot 198/0 \cdot 6411 = -1 \cdot 869$$

Then from P, $\quad x_1 = (3 \cdot 08 + 2 \cdot 39 \times 1 \cdot 869)/11 \cdot 25 = 0 \cdot 6708.$

In line 3 we have eliminated x_1 by the following method, which is always adopted:

(i) Find the numerically largest coefficient of x_1: this is called the **pivot,** and the equation in which it occurs is called the **pivotal equation** for this elimination.

(ii) To eliminate x_1 between the pivotal and another equation E, find the multiplier (M_E) by which the pivotal needs to be multiplied before subtraction from E.

M_E is always numerically less than unity: here it is

$$5 \cdot 23/11 \cdot 25 = 0 \cdot 4649 \text{ to 4S}$$

(iii) In the next line are set the coefficients of the new equation E' derived from E after this process is complete, e.g. under x_2 we have $0 \cdot 47 - 2 \cdot 39 \times 0 \cdot 4649$.

The check is that the S figure (obtained by the same process from the S-figures above it) should be again the sum of those in its line.

NOTE. We do not here justify the number of figures retained at each stage.

8.1. Solve the equations $\quad 9\cdot15x_1 + 0\cdot83x_2 = 78\cdot5$
$$0\cdot05x_1 + 2\cdot46x_2 = -36\cdot1$$

treating the given values of the coefficients as *exact*: this means that the values of x_1, x_2 can be given finally to any desired degree of accuracy (say 4S).

The S procedure checks the work as far as the equation for one variable; but error can arise in dividing to find this variable, or in the back-substitution to find the other. Substitute in the *pivotal* equation.

The complete check is given by calculating to 4S the value of the 'residual' for each equation, i.e. of

$$9\cdot15x_1 + 0\cdot83x_2 - 78\cdot5$$

and $\qquad\qquad 0\cdot05x_1 + 2\cdot46x_2 + 36\cdot1$

These should differ from zero by not more than 0·01 which represents the 4th significant figure in the numbers 78·5 and 36·1.

8.2. (a) Estimate the acceptance of the results for x_1 and x_2 in Ex. 8.1 if the coefficient of the former is in fact a rounded figure, by direct solution with **0·045** in place of **0·05**.

A set of equations in which such a process causes large changes in the calculated values is called *ill-conditioned*. An example follows.

(b) The solutions of $\dfrac{x}{7} + \dfrac{y}{6} = 2$; $5x + 6y = 71$ can be seen to be $x = 7$,

$y = 6$; and the second equation can be rewritten as $\dfrac{x}{6} + \dfrac{y}{5} = 2\frac{11}{30}$.

Consider the equations modified by rounding all the coefficients to 2 decimal places:

$$0\cdot14x + 0\cdot17y = 2\cdot00$$
$$0\cdot17x + 0\cdot20y = 2\cdot37$$

Solve these equations, treating the coefficients as exact. The pivotal multiplier for x-elimination, viz: $0\cdot14/0\cdot17$, must be rounded (say to 5S). Determine the residuals.

Notice how greatly the solutions have been affected by small changes in the coefficients: the equations are clearly ill-conditioned. Since in practice the coefficients of equations are almost always subject to error, the solutions of ill-conditioned equations are virtually useless.

[Solution note.

For 8.3.

In solving a triad of equations the pivotal procedure is adopted for each elimination. It is vital to remember that *it is always the pivotal equation which is multiplied*, but its multiplier has one value M_1 for eliminating x_1 between pivotal (P) and equation E_1 and another (M_2) for eliminating between P and E_2. Since the M's are recorded, the place to put M_1 is against E_1 and M_2 against E_2, although the M is *in each case the multiplier for P*.

The example will make this clear. The star * marks the pivotal coefficient for elimination of x_1 from the triad, and the double star ** shows the pivot for the elimination of x_2 from the ensuing pair of equations:

M	x_1	x_2	x_3		**S**
0·3	− 3	2	− 0·4	1	− 0·4
− 0·5	5	− 4	+ 7	0	8
	− 10*	1	0·1	0	− 8·9
− 0·4857	(0)	1·7	− 0·43	1	2·27 checks
	(0)	− 3·5**	7·05	0	3·55 √
		(0)	2·994	1	3·994 √
			(4S)		

$$x_3 = 0·3340$$
$$x_2 = (0 - 7·05x_3)/(-3·5) = 0·6728$$
$$x_1 = (0 - 0·1x_3 - x_2)/(-10) = 0·07060 \quad \text{[N.B. 4S not 4D]}$$

Residuals
(i) $\quad -3x_1 + 2x_2 - 0·4x_3 - 1 = \quad 0·0002$
(ii) $\quad 5x_1 - 4x_2 + 7x_3 - 0 = \quad -0·0002$
(iii) $\quad -10x_1 + x_2 + 0·1x_3 - 0 = \quad 0·0002$

Solutions, to 4S
0·07060, 0·6728, 0·3340

8.3. (a) Solve the triad of equations

$$-3 \cdot 27x_1 + 2 \cdot 05x_2 - 0 \cdot 39x_3 = p$$
$$5 \cdot 08x_1 - 4 \cdot 13x_2 + 6 \cdot 95x_3 = q$$
$$-10 \cdot 0x_1 + 1 \cdot 00x_2 + 0 \cdot 100x_3 = r$$

where $p = 1$, $q = 0$, $r = 0$.

Treat the coefficients as exact, work in 5S, and give the residuals to the largest number of figures justified by the terms which contribute to them. Finally give your set of solutions to 3S.

(b) Solve the same triad for $p = 0$, $q = 1$, $r = 0$; and
(c) again for $p = 0$, $q = 0$, $r = 1$.

It will be apparent that the pivots are the same for the three equations (so long as elimination is carried out in the same order) and the multipliers are the same. It would therefore be possible to carry out the eliminations on a single layout, with a single check-sum, as follows:

M	x_1	x_2	x_2	(a)	(b)	(c)	S
0·327	−3·27	2·05	−0·39	1·00	0	0	−0·61
−0·508	5·08	−4·13	6·95	0	1·00	0	8·90
	−10·0*	1·00	0·100	0	0	1·00	−7·90
	(0)	—	—	—	—	—	—
	(0)	—**	—	—	—	—	—
		(0)	—	—	—	—	—

Solution	Values			Residuals		
of (a)	(vii)	(iv)	(i)	(x)	(xi)	(xii)
(b)	(viii)	(v)	(ii)	(xiii)	(xiv)	(xv)
(c)	(ix)	(vi)	(iii)	(xvi)	(xvii)	(xviii)

It is convenient to insert the solutions and residuals (R) at the foot as shown, in order as they are obtained.

For 8.4.

It will be clear from Ex. 8.3 that if a triad with fixed x-coefficients is liable to appear for solution on a number of occasions with different absolute terms, then the grid of solutions obtained by your work will provide a basis for a calculation of any case. If we write the solutions of **(a)** as x_{1a}, x_{2a}, x_{3a} and so on, the solution of $p = 7$, $q = 5$, $r = 0$ is

$$x_1 = 7x_{1a} + 5x_{1b}; \quad x_2 = 7x_{2a} + 5x_{2b}; \quad x_3 = 7x_{3a} + 5x_{3b}$$

because
$$-3{\cdot}27x_{1a} + 2{\cdot}05x_{2a} - 0{\cdot}39x_{3a} = 1$$
$$-3{\cdot}27x_{1b} + 2{\cdot}05x_{2b} - 0{\cdot}39x_{3b} = 0$$
$$\therefore \ -3{\cdot}27(7x_{1a} + 5x_{1b}) + 2{\cdot}05(7x_{2a} + 5x_{2b}) - 0{\cdot}39(7x_{3a} + 5x_{3b}) = 7$$

and so on for the other equations.

A grid of nine figures like that forming the L.H.S. coefficients of 8.3 is called a 3×3 matrix. The R.H.S. coefficients also form a 3×3 matrix which is here a 'unit matrix'.

The solutions also form a 3×3 matrix. This aspect of the problem is considered again in Ex. 8.5.

After 8.4. Non-regularity for a triad.

Failure to obtain a unique solution can clearly occur if the final elimination gives $0 \cdot x_3 = d$.

If $d = 0$ *any* value of x_3 will satisfy, and if $d \neq 0$ *no* value. That there is failure depends on the left-hand coefficients; but the *type* of failure on the right-hand also. For a triad the situation can be considered in terms of coordinate geometry in three dimensions, in which an equation of the first degree represents a plane. When a unique solution is possible the equations are said to be *regular*; otherwise the equations are *non-regular*.

Immediate failure occurs if the two first equations give planes parallel (no solution) or coincident (an unlimited number of solutions).

Failure at the end of the calculation occurs if the line of intersection of the first two is parallel to the third (no solution) or lies in it (an unlimited number of solutions).

For further reading.

(See Bibliography for full titles.)

Matrices; introduction and manipulation:

Neill and Moakes, *Vectors, Matrices and Linear Equations*

Matrices in the general mathematical setting:

Moakes, *Core of Mathematics*

Linear equations:

Cohn, *Linear Equations*

Numerical methods:

Wooldridge, *Computing*.

8.4. Deduce from your triple solution of Ex. 8.3 the solutions of the same equations for the case $p = 5 \cdot 20$, $q = -1 \cdot 74$, $r = 3 \cdot 87$. Use these values as multipliers for the solutions obtained with unit right-hand sides, as shown in the note. Calculate the residuals.

8.5. Solve the two pairs of equations, using a single layout

$$5 \cdot 23x_1 + 0 \cdot 47x_2 = 1 \cdot 00 \quad \big| \quad = 0$$
$$11 \cdot 25x_1 + 2 \cdot 35x_2 = 0 \quad \big| \quad = 1 \cdot 00$$

Deduce the solution of the pair already solved on p. 48, by the method of multipliers as above.

This method is of great practical importance, and we shall set it out (for two equations only) using the algebraic notation with suffixes.

The square array ('matrix') of left-hand-side coefficients, which determines the solution procedure, will be written

$$\begin{pmatrix} a_{11} & a_{12} \\ a_{21} & a_{22} \end{pmatrix}$$

The first suffix in each case refers to the equation, upper or lower; and the second suffix to the variable, x_1 or x_2.

The solutions of the equations, solved first with $\begin{pmatrix} 1 \\ 0 \end{pmatrix}$ and then with $\begin{pmatrix} 0 \\ 1 \end{pmatrix}$ as right-hand side, will also be written in square matrix form:

$$\begin{pmatrix} \alpha_{11} & \alpha_{12} \\ \alpha_{21} & \alpha_{22} \end{pmatrix}$$

Here the first suffix refers to the variable, and the second to the solution — first of $\begin{pmatrix} 1 \\ 0 \end{pmatrix}$, then of $\begin{pmatrix} 0 \\ 1 \end{pmatrix}$ as right-hand side.

By substitution into the equations we obtain:

$$\left. \begin{aligned} a_{11}\alpha_{11} + a_{12}\alpha_{21} = 1 \\ a_{21}\alpha_{11} + a_{22}\alpha_{21} = 0 \end{aligned} \right\} \qquad \left. \begin{aligned} a_{11}\alpha_{12} + a_{12}\alpha_{22} = 0 \\ a_{21}\alpha_{12} + a_{22}\alpha_{22} = 1 \end{aligned} \right\}$$

These are the conditions that the product of the a-matrix into the α-matrix is the unit-matrix $\begin{pmatrix} 1 & 0 \\ 0 & 1 \end{pmatrix}$; i.e. that the a- and α-matrices are inverse. The process of obtaining the α's is called *inverting the matrix of coefficients*. We have seen in Ex. 8.4 that the result enables us to derive the solution of any set of equations with these coefficients, whatever the right-hand sides may be.

(See references on left-hand page, for further reading.)

For 8.6.

Solving linear equations by successive approximations.

We have seen that the acceptability of solutions can only be shown by slightly varying them and watching the effect on the residuals. The following method follows this approach, starting with arbitrary values of the variables.

Consider the equations
$$2x + 5y - 8 = R_1;$$
$$30x + y + 28 = R_2$$
where we require x, y such that $R_1 = R_2 = 0$.

We start with a schema, in which the coefficients figure in a new role (and *rearranged*), viz: to determine the changes ΔR_1, ΔR_2 made by changes $\Delta x = 1$, $\Delta y = 1$ in the values of the variables. As a check we calculate at every stage also the value of $R = R_1 + R_2$. Our schema is

			ΔR_1	ΔR_2	ΔR
(a)	$\Delta x = 1$		2	30	32
(b)		$\Delta y = 1$	5	1	6

Note that line **(a)** contains the first *column* of coefficients, and line **(b)** the second column, from the given equations.

We now start with, say, $x = 0$ and $y = 0$ and trace the solution through: at each stage we reduce one of the residuals to zero or conveniently near zero.

					R_1	R_2	R
Initially	$x = 0$		$y = 0$		-8*	28	20
	Δx		Δy				
By **(b)**			1·6		0	29·6*	29·6 ✓
(a)	-0.99				-1.98*	-0.1	-2.08 ✓
(b)			0·4		0·02	0·3*	0·32 ✓
(a)	-0.01				0	0	0 ✓
Solution	—1·00		2·00				

* The starred figure in each case controls the next move.

The operative values of Δx, Δy have been chosen as follows:

(i) To add 8, to make R_1 zero, $\Delta y = 8/5$.

(ii) To reduce $R_2 = 29.6$ to zero *exactly* we should take $\Delta x = (-29.6)/30$, but we prefer at this stage the 2D value $\Delta x = -0.99$.

(iii) Again, as in (ii), we shall not allow a complicated value of a change to be used. Take $\Delta y = 0.4$ *exactly*, to bring R_1 nearly to zero.

8.6. Repeat the example given, making as the first step a change in x to reduce R_2. This is in accordance with the working rule, *not followed in the example*, of reducing first the largest residual. Notice that it is not necessary to reduce R_2 to *zero*. Progress is made, without introducing heavy figures at an early stage, if we can reduce the magnitude of R_2 by a factor of about 10: thus here it would be quite enough to reduce R_2 from 28 to -2.

The strict method of solution, in which the R's are made to vanish alternately, can be seen by plotting graphs of the loci to be 'spiralling' round the point of intersection if the gradients are of opposite sign, switching from one line to the other and back:

$x = OA,$
$\quad y = 0$

is a pair of trial values.

OA, AB is the next pair. The coords of C, the next; then of D, and so on.

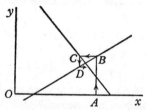

When R_2 is made to vanish we obtain a pair of coordinates which satisfy line 1 only: one coordinate is carried over to obtain a point on line 2, and so on.

If, however the residuals are only made approximately zero, one is working with a *band* rather than a line. In the end stages the residual *must* be made zero.

It will have been seen that the speed of this method of solution depends a great deal on judgment in choosing which residual to reduce first and by how much. There is nothing to prevent the solver from building up modifications in any way — including a change of two or more values at once — and freed from its mechanical character this is known as the method of **Relaxation**. See Note 8.8.

For computer work, on the other hand, the purely mechanical routine of the spiral is appropriate; but the order of operations must be chosen in such a way as to ensure that the solution converges. (It may do so by steps, not by a spiral.)

The same method is available when one or more of the equations is non-linear: in fact there may be no alternative in such cases.

For further reading see Wooldridge pp. 115 *et seq.*

For 8.7. If the absolute term had been 28·1 in place of 28, the columns R_2 and R would have been different, and after the fourth line $(R_2 = 0.4^*$ $R = 0.42)$ we should multiply temporarily by 1000 and proceed as follows

		400	20	400*	420
By (a)	-13		-6^*	10	4 √
		1·2	0	11·2*	11·2 √
(b)	-0.37		-0.74^*	0·1	-0.64 √
		0·15	0·01	0·25	0·26 √
(a)	-0.01		-0.01	-0.05	-0.06 √

For 8.8.

Given the relaxation scheme

		ΔR_1	ΔR_2	ΔR_3	ΔR
(a)	$\Delta x = 1$	-5	2	-3	-6
(b)	$\Delta y = 1$	6	7	8	21
(c)	$\Delta z = 1$	-1	4	2	5

before going on we form combinations which will relax two or if possible only one residual at a time, e.g.

			ΔR_1	ΔR_2	ΔR_3	ΔR
(d)	$\Delta x = 1$ $\Delta y = 1$	$\Delta z = 1$	0	13	7	20 √
(e)	$\Delta y = -1$	$\Delta z = 4$	-10	9	0	-1 √
(f)	$\Delta x = 2$	$\Delta z = -1$	-9	0	-8	-17 √

		Then if the data give, for:	R_1	R_2	R_1	R
	$x = 0$　$y = 0$　$z = 0$		1	-9	-8	-16
By f:	$\Delta x = -2$	$\Delta z = +1$	10	-9	0	1 √
By e:	$\Delta y = -1, \Delta z = 4$		0	0	0	0 √

and the solution is

$$x = -2, y = -1, z = 5.$$

For 8.9.

If however the residuals R_1 etc. above were 1, -8, -8 the final step would give

　　　　　　　　　　　0　　1　　0　　1

We have *two* residuals zero, but no combination which reduces this. We obtain it thus: combine d and f to get *another* set ending in zero (it is $8d + 7f$).

(g)	$\Delta x = 22, \Delta y = 8, \Delta z = 1$	-63	104	0	41 √
	This minus (e) $\times 6.3$ is	0	47·3	0	47·3 √
	i.e. $\Delta x = 22, \Delta y = 14.3, \Delta z = -24.2$.				

and this final modification $\times (-1/47.3)$ gives the required solution.

8.7. Verify that the mechanical process on the left hand page gives $x = -1 \cdot 00338$ $y = 2 \cdot 00135$ and show by direct substitution that the residues given are correct. Repeat Ex. 8.6 with **28·1** for 28 and continue, to get a solution with residuals less than 10^{-3}.

8.8.

(i) State the triad of equations for which the given scheme is appropriate, and check directly that the solution satisfies them.

[Solution note.

(ii) Solve the equations with the same x-coefficients but for which the residuals (for zero x, y, z) are -10, -4, -7.

(iii) Solve the equations

$$x + y + z = 0 \cdot 9$$
$$10x - y + 13z = 1 \cdot 0$$
$$7x + 2y - 3z = 3 \cdot 8$$

8.9.

(i) Solve, to the accuracy required to obtain residuals numerically less than 10^{-2}

$$-5x_1 + 6x_2 - x_3 = 10$$
$$2x_1 + 7x_2 + 4x_3 = 5$$
$$-3x_1 + 8x_2 + 2x_3 = 8$$

(ii) Solve, with residuals numerically less than 10^{-3}:

$$x_1 + x_2 + x_3 = 1 \cdot 0$$
$$10x_1 - x_2 + 13x_3 = 1 \cdot 0$$
$$7x_1 + 2x_2 - 3x_3 = 4 \cdot 1$$

9

NUMERICAL INTEGRATION

This process is equivalent to finding the area 'under' a curve $y = f(x)$ between given ordinates, by calculation of a series of ordinates at intervals throughout the range; e.g. to find $\int_0^a f(x)\,dx$ by dividing the range 0 to a of x into a number of parts, usually $2n$ of width h so that $a = 2nh$.

Elementary methods are (i) the trapezium rule, by which each area of width $2h$ is treated as a trapezium. The first will have area

$$\tfrac{1}{2}\{f(0) + f(2h)\}2h;$$

and so on. (ii) Alternatively we could take the mid-ordinate as the effective height of each $2h$-strip; and give the approximate area as $f(h) \times 2h$ for the first and so on through the ordinates at odd multiples of h.

A little examination of curves will show that when one result is too small the other is too large and vice versa, and the method most used in practice combines the two (Simpson's rule).

Simpson's Rule states that a good approximation to the area of the first $2h$-strip is $\qquad (h/3)\{f(0) + 4f(h) + f(2h)\}$

This is equivalent to writing

$$\int_0^{2h} f(x)\,dx \approx \frac{h}{3}\{f(0) + 4f(h) + f(2h)\} \qquad\qquad\text{(S)}$$

Similarly

$$\int_{2h}^{4h} f(x)\,dx \approx \frac{h}{3}\{f(2h) + 4f(3h) + f(4h)\}$$

and so on, for each of the n intervals each of width $2h$.

Our procedure therefore with the $2n+1$ tabulated values of f is to form a weighted sum by multiplying them respectively by 1, 4, 2, 4, 2, ... 4, 1. On dividing this by the sum of the weights (which should be $6n$) we obtain the *Simpson mean ordinate*. Finally this mean, multiplied by the range of integration, gives the result.

9.1.

Determine by Simpson's rule an approximate value of

$$\int_0^1 \frac{dx}{1+x},$$

taking **h = 0·1**. Employ the layout below:

x	1 + x	f(x) values	multipliers	
0	1	1·000,00	1	
0·1	1·1	0·909,09	4	(Space for differ-
0·2	1·2	0·833,33	2	ences if used)
0·3	1·3	0·769,23	4	
		etc., to:		
1·0	2·0	0·500,00	1	

Determine the cumulative sum Σ, checking that the cycle register shows the total 30. Why?

Then the value required is $\Sigma/30$, since the range of integration is unity.

Compare the result, rounded to 4S, with the true value of $\log_e 2$.

9.2. Repeat the above calculation with **h = 0·05**.

9.3. With $h = 0·1$ find a value for

$$\int_0^1 \frac{dx}{1+x^2} \text{ to 4S; working to 6D}$$

and compare the result with the known value of $\pi/4$ to 4S.

Note: the routine, of carrying two places more than are finally required, does not necessarily secure an accurate result *when an approximate formula is used.* The *adequacy* of the method, with any correcting devices included, has also to be considered.

The $f(x)$ values will be needed again (Ex. 9.6).

9.4. Find a value of $\displaystyle\int_0^{0·8} \frac{dx}{1+x^3}$, **working to 5D**

(i) with **h = 0·2** and (ii) with **h = 0·1**.

It will be advisable to form the third column with *all values of $f(x)$ for intervals 0·1*, and to extract the required values for the two calculations separately.

The comparison of the two results will give an indication of how many figures in the result can be relied on.

For 9.5.

It will be seen that if the integral is one which can only be evaluated numerically — and it is for these that the method is most useful — one might have to face a series of redeterminations in order to get the required accuracy.

This is the method used on an electronic computer, but for desk work we must know for any chosen value of h the order of error to be expected, and if possible a correction which we can apply.

We start with the fact that the basic equation (S) of p. 58 is *exact* if $f(x)$ is a polynomial in x of degree 3 or less. In verifying this it is easier to take the origin at the centre of the range of x, so that we prove:

$$\int_{-h}^{h} f(x)\, dx = \frac{h}{3} \{ f(-h) + 4f(0) + f(h) \}$$

The reader should write $f(x) = a_0 + a_1 x + a_2 x^2 + a_3 x^3 + px^4$ and establish that there is a discrepancy proportional to ph^5. This is zero if p is zero, and can otherwise be reduced to acceptable size by diminishing the interval h.

Looking at the Simpson method for a range $a = 2nh$, we see that if $f(x)$ can be represented well enough by polynomials of degree 4 — they can be different polynomials for each range $2h$ — then the correction to be applied is

$$C = -\tfrac{4}{15} h^5\, n\bar{p}$$

where $\bar{p}$ is the mean of the values of p.

We have seen that the 4th differences $\Delta^4 f$ for such a polynomial function will be $(4!)ph^4$. We may therefore rewrite the correction as

$$C = -\tfrac{1}{90} hn\mathscr{D}$$

where $\mathscr{D}$ is the mean fourth difference.

$$= -a\mathscr{D}/180$$

since $a = 2nh$.

A value for $\mathscr{D}$ can be assigned sufficiently well from the $(2n-4)$ values of the 4th differences which will arise from the calculated values of f; and which will have been determined already in order (i) to check the accuracy of the f values; and (ii) to establish the adequacy of this method, for the chosen value of h.

9.5.

Form a difference table for Ex. 9.1 and compute the correction: compare the corrected value with the true value.

Estimate, without recourse to a new table, the correction to be applied in Ex. 9.2. The 4th differences, being of the order of h^4, will be approximately $\frac{1}{16}$ of the previous values; but there will be twice as many.

9.6. Tabulate the function of Ex. 9.3 and apply the correction calculated from 4th differences.

Notice that for small x, $1/(1+x^2) \approx 1 - x^2 + x^4$ so that the earlier $\Delta^4 f$ should be $24h^4$; but this expansion is only appropriate near $x = 0$. Near $x = 1$ write $x = 1 - z$ and we have

$$1/(1+x^2) = \tfrac{1}{2}\{1 - z - z^2/2\}^{-1}$$
$$= \tfrac{1}{2}\{1 + (z - z^2/2) + (z - z^2/2)^2 + (\quad)^3 + (\quad)^4 + ...\}$$

Examination of the term in z^4 shows differences to be $6h^4$ in this part of the table.

9.7. Use the tabulated values of Ex. 9.4 to form a table of

$$f(x) = 1/\sqrt{(1+x^3)}$$

for the same range.

Use an iterative method for this square root starting with the value from 4-figure tables; or as in note below.

Form 4th differences, correct your Simpson result and indicate to how many figures you consider your result reliable.

Your check can be *either* two corrected Simpson results for different h; *or* a different integration formula.

Note: Working directly from values of $N = (1 + x^3)$ you could take one move only to find $N^{-1/2}$. Write $F(x) \equiv (1/x^2) - N = 0$.

Verify that Newton's rule gives for this case

$$x_{n+1} = \frac{x_n}{2}(3 - Nx_n^2)$$

10

NUMERICAL DIFFERENTIATION

Consideration of the graph of a function shows that *integration* will give a result to rather *greater* % accuracy than the worst of the data: e.g. if a function is represented to 1% or less over the whole range, then the Simpson formula will generally give a result well within 1%.

On the other hand, *differentation* gives a result with *smaller* accuracy. This will be familiar to a physicist who has drawn a cooling curve for a body by plotting temperature against time, and has tried to derive from it a series of values of the rate of fall of temperature. He finds that concordant results can only be found by 'smoothing' the data. A chord drawn to join successive points will indicate by its gradient a good approximation to the rate of fall at the middle of the time-interval, *provided* the data are known to a high enough accuracy to give a smooth curve.

The criterion of smoothness for a series of numbers is the regularity of the differences. However, nth differences can always be irregular to the extent of $(1/2) \times 2^n$, since the final figures of the data are uncertain to the extent $\pm (1/2)$; differentiation therefore requires say 3 extra figures in the data in place of the usual 2, to get a result to a required number of figures.

We use as a first approximation

$$f'(x) \approx (1/h)\{f(x+h/2) - f(x-h/2)\}$$

or $$f'(x+h/2) \approx (1/h)\{f(x+h) - f(x)\} = (1/h)\Delta f(x)$$

If a difference is regarded as *belonging to the value of the function opposite which it is written* the symbol δ is used, and we write

$$f'(x) \approx (1/h)\delta f(x) \dots\dots\dots\dots\dots\dots\dots (10.A)$$

Notice that if f is tabulated for $x = 0$, 0.2, 0.4 etc. up to 1.0 the differences lie opposite the missing values 0.1 etc. and the formula gives (approximately) $f'(0.1)$, $f'(0.3)$ to $f'(0.9)$.

Consideration of polynomials gives the improved formula

$$f'(x) \approx (1/h)\{\delta f - (1/24) \delta^3 f + (3/640) \delta^5 f\} \dots\dots\dots\dots (10.B)$$

in which omitting the last term is less serious usually than errors due to the data, the values of f.

10.1. Referring to $f(x) = \dfrac{1}{1+x}$, tabulated and differenced in Ex. 9.5 for

0 (0·2) 1·0, obtain values of $f'(x)$ for 0·3 (0·2) 0·7

    (i) from a graph, using 3S for $f(x)$;

    (ii) using formula (**10.A**) or the simpler forms given above it;

    (iii) using (**10.B**), remembering that $\delta^3 f$ means the value on the same line as the value of x being considered;

and (iv) by calculating $f'(x) = -1/(1+x)^2$.

10.2. If $f(x) = a_0 + a_1 x + a_2 x^2 + a_3 x^3 + a_4 x^4$, show that
$$f(k) - f(-k) = \delta f(0) = 2a_1 k + 2a_3 k^3$$
(We are taking $2k = h$, the tabular interval.)

   (a) Hence show that the error in (**10.A**) is $a_3 k^2$ or $a_3 h^2/4$.

   (b) Find $\delta f(k) = f(2k) - f(0)$ ⎫ and hence evaluate
   and $\delta f(-k) = f(0) - f(-k)$ ⎭ $\delta^2 f(0) = \delta f(k) - \delta f(-k)$.

   Show that $f''(0) \approx (1/h^2)\delta^2 f(0)$ with an error of order $a_4 h^2$. (Notice that $k = h/2$ is only used as a matter of notation: no intermediate *tabular* values in the h-intervals are used.)

   (c) Using 4-figure values of $\sin x$ over 0°(10°) 90°, form a difference table. Obtain values of $f''(x)$ for x in *radians* by writing $h = \pi/18$ in the equation
$$f''(x) \approx (1/h^2)\delta^2 f(x) \dots\dots\dots\dots\dots\dots(10.C)$$
and compare with values of $-f(x)$.

10.3. Form a table of $\log_{10} x$ to 6D, $x = 1(1)10$.

Difference and find $f'(x)$ for 2·5 (1) 8·5 by (10.B) and compare with values of $(·4343)/x$.

For 10.4. This example was important because if f could be represented by a formula it would be possible to predict with some confidence the behaviour of the substance *beyond* the range of the experiments. The design of power plant depends on this function for steam.

For 10.5.

Formula (10.C) can be rewritten in the form

$$h^2 f''(x) \approx \delta^2 f(x) = \{f(x+h) - f(x)\} - \{f(x) - f(x-h)\}$$
$$= f(x+h) - 2f(x) + f(x-h) \dots\dots\dots\dots(10.D)$$

Thus if we are *given* a rule by which $f''(x)$ can be evaluated for any x (10.D) gives a method by which the values of $f(x-h)$ and $f(x)$ can be used to find a value of $f(x+h)$, with an error of order h^4. This is one of several methods for numerical solution of differential equations of the second order. (There are devices by which the correction terms of order h^4 are fed in; either on the way, or en masse after tabulation of the approximate solution.) E.g. if

$$f''(x) = -f(x)$$

then **10(D)** becomes

$$f(x+h) \approx (2 - h^2)f(x) - f(x-h)$$

Layout as follows:

x	$x+h$	$f(x-h)$	$f(x)$	$f(x+h)$ approx.
		(1) given; and	(2) given→	(3) calculated
—		(2) given	(3) from above,→	(4)

and so on.

The full set of values, after the given pair, reads down the final column, the values of the variable being those in the *second* column.

(The arrows represent work which in some cases might require a further column in the layout.)

10.4. 'Live' examples of numerical differentiation are rather rare. The following is one from physical chemistry.

The energy change involved in the dissociation of a molecule of a certain compound is a function $f(T)$ of the temperature (°C) at which dissociation occurs. This function is required, and can only be found by determining experimentally another function $q(T)$ which is known to be connected with it by the equation

$$A \frac{dq}{dT} = \frac{f(T)}{(T+273)^2}$$

You are given the smoothed experimental data as follows:

T	5	10	15	20	25	30	35	40	45	50
q	1·700	1·962	2·175	2·347	2·481	2·581	2·652	2·700	2·733	2·767

Taking $A = 10^{-5}$, form a table of f for $T = 12·5(5)$ 42·5 and by differencing this table state whether f could be represented by a polynomial over this range of values.   [Solution note.

10.5. Solve approximately the equation $f''(x) = -f(x)$ to find values of $f(x)$ at intervals of 0·1 in x given the initial values $f(0) = 0$ and

$$f(-0·1) = -0·099833$$

using the relation, applicable to this equation only

$$f(x+h) \approx (2-h^2)f(x) - f(x-h)$$

Errors are accumulating, and it will be possible to see at what point the value of $f(x)$ differs from $\sin x$ in the *4th* place.

10.6. Using the expansion

$$\cos(h) \approx 1 - h^2/2 + h^4/24 - h^6/720$$

show that, to order h^4,

$$\cos(x+h) = \cos x\{2 - h^2 + h^4/12\} - \cos(x-h)$$

and hence repeat the process of Ex. 10.5 with the correction 'built in', in the revised multiplier.

Start with $f(0) = 1$ and $f(h) = f(0·1)$ evaluated by the expansion. Work with all the figures your machine allows, up to $\cos(1·5) = 0·070737$ to 6D.

REVISION EXERCISES AND HARDER PROBLEMS

Exercise A. Simultaneous linear equations.

A.1. Solve by pivotal condensation with a sum-check the simultaneous equations:

$$2{\cdot}598x_1 - 1{\cdot}500x_2 = 0{\cdot}675$$
$$1{\cdot}449x_1 + 0{\cdot}388x_2 = 1{\cdot}500$$

Work to 5 decimal places throughout, and compute the residuals also to 5 places.

A.2. The unknown x and y are given by the simultaneous equations:

$$1{\cdot}08x + 2{\cdot}56y = 1{\cdot}96$$
$$-0{\cdot}49x + 1{\cdot}82y = 0{\cdot}97$$

where the coefficients may be treated as exact.

(a) Obtain approximate solutions to 2S using a slide-rule. A single setting will multiply the coefficients of the first equation so as to give $0{\cdot}49$ as the coefficient of x: then by addition a value of y is obtained, and so on.

(b) Verify by machine that to 5D, $y = 0{\cdot}62360$: obtain the value of x and the residuals to the same number of places.

A.3. Show how it is possible with a single layout to solve the simultaneous equations

$$0{\cdot}4000x + 0{\cdot}1000y = a$$
$$0{\cdot}2410x + 0{\cdot}6000y = b$$

both for the case $a = 1$, $b = 0$ *and* for $a = 0$, $b = 1$. Carry out the solution to 4D and verify by direct calculation that the four results, in a suitable array, form the inverse of the matrix $\begin{pmatrix} 0{\cdot}4 & 0{\cdot}1 \\ 0{\cdot}241 & 0{\cdot}6 \end{pmatrix}$.

A.4. Solve, working to 6 significant figures in the coefficients,

$$-1{\cdot}69843a + b = -0{\cdot}0068540$$
$$a - 1{\cdot}67836b = -0{\cdot}020562$$

(This arose from a 3-step solution of a differential equation. The first and fourth values of $f(x)$ were given; a, b are the intermediate values.)

A.5. Explain briefly the meaning of the term ill-conditioned when applied to a system of simultaneous linear equations.

Solve the following (well-conditioned) simultaneous equations by the method of pivotal condensation with a sum check. Give your answers correct to three significant figures.

$$6x + 3y - z = 1$$
$$x - 5y + 2z = 11$$
$$x - y + 8z = 17$$

(M.E.I.)

A.6. Solve, carrying 5 decimal places in your working, the set of equations:

$$2 \cdot 731x_1 + 5 \cdot 826x_2 + 1 \cdot 743x_3 = 3 \cdot 000$$
$$3 \cdot 285x_1 + 1 \cdot 721x_2 + 3 \cdot 618x_3 = 2 \cdot 000$$
$$2 \cdot 153x_1 + 4 \cdot 876x_2 + 5 \cdot 232x_3 = 1 \cdot 000$$

Give your results to 4 decimal places, together with the corresponding residuals.

Exercise B. Polynomials, including *nesting* and the use of *differences*. Solution of polynomial equations using *linear interpolation* and *Newton's method*.

Notes.

(1) In using *nesting* on a machine (as distinct from a computer program) it may often be worth while to modify the procedure to put off change of sign to the last step. A simple example is in question **B.2** where it is helpful to tabulate first the positive function $g(x)$ which is $f(x)$ without its last term.

(2) When using differences to detect errors in a table it is advisable to work right through the columns from each tabular value in turn. In this way each anomaly can be detected as it occurs, and after it has been corrected any subsequent error can be picked up just as easily. Question **B.4** illustrates the value of this.

B.1. (a) The following table contains an error. By constructing a difference table, locate the error and correct the original table.

x	1	2	3	4	5	6	7	8	9	10
y	7	12	21	34	51	70	97	126	159	196

(**b**) Show that the equation $x^3 - 3x + 1 = 0$ has a root lying between 0.3 and 0.4. **Either** find this root correct to 4 significant figures **or** write a flow diagram for Newton's or some other iterative method for calculation by a digital computer of this root correct to six significant figures.

(M.E.I.)

B.2. Evaluate the polynomial

$$f(x) = x^3 + 2.910x^2 + 4.131x - 32.75$$

for the values $x = 0, 1, 2, 3, 4$.

Obtain by linear interpolation an approximation (expressed to two decimal places) to a root of $f(x) = 0$ in this range. Calculate by any method the value of this root to an accuracy of 4 significant figures and write a simple flow diagram for the method which you use.   (M.E.I.)

B.3. Prove that when a polynomial $f(x)$ of degree n is tabulated for equally spaced valued of x, the nth differences are equal. Make use of this fact to tabulate:

$$f(x) = 2.31x^3 + 3.12x^2 - 4.72 \text{ for } x = -2(0.5)2.$$

Find correct to 3 significant figures, the real root of the equation $f(x) = 0$.   (M.E.I.)

B.4. The following values are given for a polynomial $f(x)$ of the third degree, from $x = 0$ to $x = 1$ at equal intervals of 0.1 in x:

$$1.000, \ 0.991, \ 0.968, \ 0.937, \ 0.904, \ 0.875,$$
$$0.865, \ 0.853, \ 0.872, \ 0.991, \ 1.000$$

Discover whether they contain any errors and, if so, correct them. Obtain by linear interpolation an approximate value for $f(0.46)$. State, giving your reasons, whether the value so obtained is above or below the true value.   (Lond.T.Cert.)

B.5. The differences of a polynomial $f(x)$ are formed from its values at $x_{-3}, x_{-2}, x_{-1}, x_0, x_1, x_2$ and x_3. Show, by calculating the fourth differences, how an error of ε in the value of $f(x_0)$ will affect the table of differences of the function. Estimate the maximum possible error in the fifth differences if ε is a rounding off error in the least significant digit.

Tables of values for two quartic [cubic] polynomials $g(x)$ and $h(x)$ are given, to three significant figures, as

x	1·0	1·1	1·2	1·3	1·4	1·5	1·6
g(x)	1·00	1·13	1·35	1·76	2·10	2·69	3·46
h(x)	4·00	4·87	5·91	7·15	8·60	10·3	12·3

Correct any errors other than those attributable to rounding off, in either table. (M.E.I.)

B.6. (a) Write down the first three orders of differences for the following table of values. Locate and correct a probable error in the table and suggest how this type of error may occur.

x	7	8	9	10
y	103·32	110·62	118·04	125·62

x	11	12	13	14
y	133·04	141·42	149·72	158·34

(b) If the variable y in **(a)** is a cubic polynomial in x, use the table of differences to express y in the form:

$$a + b(x-7) + c(x-7)(x-8) + d(x-7)(x-8)(x-9)$$

Hence evaluate y to 2 decimal places when $x = 9·4$, and compare the value obtained from the cubic with that obtained by linear interpolation. (M.E.I.)

B.7. Calculate the values of $x^3 - 3·7x^2 + 2·1x + 1·2$ for $x = -2(1)5$ using a method in which you evaluate initially as few values as are necessary to form an appropriate difference table and then build up the remaining values.

Use the table, with linear interpolation, to find approximately the roots of the equation $f(x) = 0$. For each such solution state (with reasons) whether it is above or below the true value. (Lond.T.Cert.)

Note. Questions B6 and B7 were set in papers for which machines were not provided. For machine solution the accuracy requirements can be made more exacting.

B.8. Show that $x^3 + 3x - 1 = 0$ has just one real root and evaluate it by any method to 2 significant figures.

Explain with the aid of a flow-diagram how you would use Newton's method to find the root correct to 8 decimal places. (M.E.I.)

B.9. Prove that the values of **k** for which the set of equations

$$
\begin{aligned}
x_1 \quad\quad + x_3 &= kx_1 \\
-2x_1 + 2x_2 \quad\quad &= kx_2 \\
-x_1 + \quad x_2 - x_3 &= kx_3
\end{aligned}
$$

has solutions other than $x_1 = x_2 = x_3 = 0$ satisfy the equation $k^3 - 2k^2 + 2 = 0$. Prove that one root of this equation lies between -1 and 0 and use an iterative method to find the root to 2 decimal places. (M.E.I.)

B.10. Consider the iterative process $x_{n+1} = \frac{1}{4}x_n(5 - Nx_n^4)$. Show that if it converges it does so to $N^{-\frac{1}{4}}$. Hence or otherwise find $(3)^{-\frac{1}{4}}$ and $(27)^{-\frac{1}{4}}$ to 6 significant figures. Use these results to check one another.

B.11. Show that the equation $y^3 - 3y = 0.90$ has one positive and two negative roots. Obtain the positive root by Newton's method to 4 significant figures and show on a flow-chart how the process could be continued to 6 significant figures (supposing that the accuracy of the given coefficients justifies this). (M.E.I.)

B.12. Give a clear graphical explanation of Newton's iterative formula $x_{r+1} = x_r - f(x_r)/f'(x_r)$ for a root of the equation $f(x) = 0$. Hence derive an iterative formula for $\sqrt{(1/N)}$ and use it to obtain $\sqrt{\frac{1}{3}}$ correct to 6 decimal places. Use tables for your first approximation.

Exercise C. Solution of equations involving tabulated functions.

A flow-chart made for any of these exercises will involve using a table or, if it is translated into a computer program, a subroutine for computing the required function. (See Appendix IV, on function tables.)

 C.1. Use Newton's method to find a solution of the equation:

$$\sin x + x = 0.8$$

as accurately as your tables permit. (If required, take 1 radian $= 57.296$ degrees.)

C.2. Find by inspection of the tables an approximate root in the range $1 < x < 6$ of the equation $e^x - 4x = 0$. Use Newton's method to get this root correct to 3 decimal places.

Note that for some forms of table for e^x the intermediate values are best obtained by multiplication, e.g. $e^{2 \cdot 502} = e^{2 \cdot 5} \times e^{0 \cdot 002}$. Take $e^{0 \cdot 01}$ as $1 \cdot 01005$ and $e^{0 \cdot 001}$ as $1 \cdot 00100$.

C.3. Show by a sketch that there are three solutions of $\sin x = \frac{3}{4}x$, where x is in radians, and that one of these is positive. Find this solution to 4 significant figures.

C.4. Show graphically or otherwise that there is only one positive root of the equation:

$$e^x = 1 \cdot 100 + 2 \cdot 005x$$

and find it to 3D by any method.

C.5. Sketch on the same axes the graphs of $y = x^2 - 1$ and $y = \sin\frac{1}{2}x$: hence show that there is a root of $x^2 - 1 = \sin\frac{1}{2}x$ between 1 and 1·5. Use an iterative method to find this root to 4 significant figures.

(Lond.T.Cert.)

C.6. It is required to tabulate the root, lying between 1 and 10, of the equation $1 + \ln x = kx$ for $k = 0 \cdot 5(0 \cdot 1)0 \cdot 9$. Using any method or combination of methods, form such a table. More credit will be given for a complete table of moderate accuracy than for fewer results of higher accuracy. The accuracy claimed should in any case be stated and substantiated.

(M.E.I.)

C.7. Sketch using the same axes and scales the graphs of $y = 3 \log_e(x + 2)$ and $y = x^2$. From your graphs read off approximately a positive solution of $x^2 = 3 \log_e(x + 2)$. By successive approximation obtain this root to 4 significant figures, indicating clearly how you proceed from one approximation to the next.

(Lond.T.Cert.)

C.8. (a) Solutions correct to 4 decimal places of the equation $\log_{10}(1 + x) = c - \dfrac{x}{10}$ are listed for $c = 0 \cdot 20 \ (0 \cdot 05)0 \cdot 70$ as follows:

0·4344, 0·5623, 0·6987, 0·8435, 0·9968
1·1585, 1·3288, 1·5057, 1·6948, 1·8904, 2·0943

There is an error: use a difference method to find and correct it.
(b) Use your corrected table to solve approximately

$$10 \log_{10} y = 4 \cdot 25 - y$$

(Lond.T.Cert.)

Exercise D. Flow-charts and computers.

If the student has access to a computer he may wish to translate these flow-charts into a form acceptable by it. Some examples of such translation are given in Moakes, Croome and Phillips, *Pattern and Power of Mathematics* Book 7 (Macmillan).

Note. The first two questions were framed for solution without a machine.

D.1. By sketching the graphs of $y=x$ and $y=5-\dfrac{5}{x^2}$ for an appropriate range of values of x find to two significant figures the positive roots of the equation:

$$x=5-\frac{5}{x^2}$$

The iterative method shown in the flow-chart below can be used to evaluate only *one* of these roots; find by trial (with the help of your graph or otherwise) which root this is and determine it as accurately as your 4-figure tables permit. State the precise form of question you would put in the decision box if, working with a machine, you aimed at an accuracy of 10^{-4}.

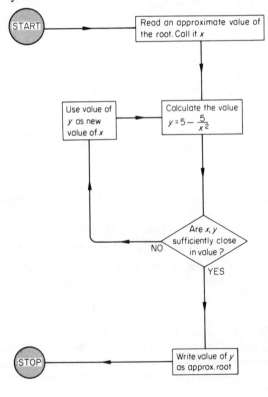

(M.E.I.)

D.2. The flow chart shown is for the solution of a certain equation with given coefficients. Correct the error in one instruction and carry out the solution for each set of data below, giving results as accurately as your tables permit.

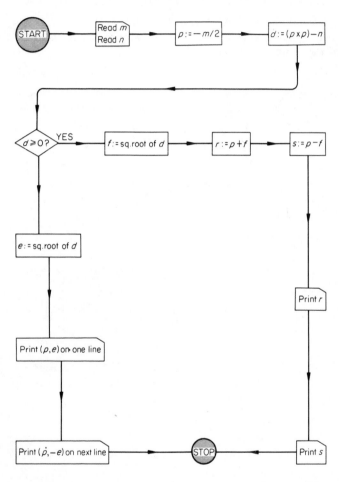

Data (read off in pairs *m*, *n*) (i) −4, 4·25; (ii) −4, 0·08

The data are treated as exact. If any result is not reliable to 3 significant figures use and explain a modified method which determines it to this accuracy. (M.E.I.)

D.3. (a) We denote by I_k the value $\int_0^{\frac{\pi}{2}} \sin^k x \, dx$ where **k** is any integer $\geqslant 0$. By differentiation of $\sin^{k-1} x \cos x$ or otherwise show that

$$I_k = \frac{k-1}{k} I_{k-2}$$

(b) Examine the given flow-chart for evaluating I_k when **k** is odd, and correct it if necessary. Supply the missing section to cover the cases in which **k** is not odd.

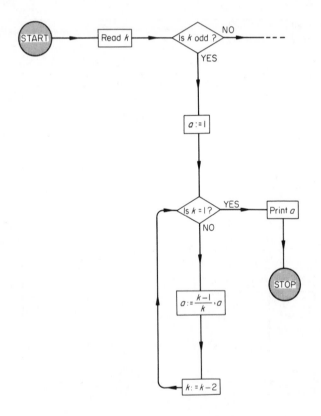

(c) Evaluate $\int_0^{\frac{\pi}{2}} \sin^4 x \cos^2 x \, dx$, $\int_0^{\frac{\pi}{2}} \sin^2 x \cos^4 x \, dx$

(M.E.I., S paper)

Computers, elementary (Ref. Lovis, *Computers I* and *II*, Arnold)

D.4. Write a brief essay on methods of input and output used in electronic digital computers, mentioning any advantages, disadvantages or special uses associated with the various types of equipment.

(St. Dunstan's O-level).

Computers, advanced

D.5. (a) Describe any standard convention for the storage of a floating point number in two storage registers each having six binary digits, illustrating your answer by showing how the decimal numbers **44, −0·0375** and **0** would be stored.

(b) A computer has binary stores named **1, 2, 3, 4** and an accumulator **A**. The following instructions are available:

Instruction code	Meaning
Set (**P**)	Place contents of store **P** into **A**.
Add (**P, Q**)	Add contents of store **P** to contents of **A** and place result in store **Q** leaving **A** empty.
SR	Shift contents of **A** one binary place to the right.
SL	Shift contents of **A** one binary place to the left.

Initially a number **n** is in store 1 and a zero in store 2. Write a program to place 10**n** in store 3. (Any number is stored having its least significant digit on the right.)   (M.E.I., S paper, Appl.Math.)

Exercise E. Numerical Integration

E.1. The values are tabulated below, to 5D for **t=0(0·2)2·0,** of the function $\phi(t) = k \exp[-\tfrac{1}{2}t^2]$ where $k = \dfrac{1}{\sqrt{(2\pi)}}$

(a) Working to 5D throughout, use the table to evaluate

$$\int_0^2 \phi(t)\, dt, \text{ and deduce } \int_{-2}^2 \phi(t)\, dt.$$

(b) What feature derived from the table shows that the function has an inflexion at or near **t=1**?

t	0	0·2	0·4	0·6	0·8	1·0
$\phi(t)$	0·39894	0·39104	0·36827	0·33322	0·28969	0·24197

t		1·2	1·4	1·6	1·8	2·0
$\phi(t)$		0·19419	0·14973	0·11092	0·07895	0·05399

E.2. Tabulate the function $g(x) = \sqrt{1+x+x^2}$ to 5S for $x = 0(0 \cdot 1)0 \cdot 8$. Obtain values for $\int_0^{0 \cdot 8} g(x)\, dx$ by Simpson's rule using (a) 5 ordinates, (b) 9 ordinates. Compare the values and comment.

E.3.

(i) Prove that if $f(x)$ is the quartic polynomial:

$$f(x) = a + bx + cx^2 + dx^3 + ex^4$$

then $\displaystyle\int_{-h}^{h} f(x)\, dx - \tfrac{1}{3}h\{f(-h) + 4f(0) + f(h)\} = \frac{-4eh^5}{15}$

(ii) By dividing the interval $(0, 1)$ into four equal parts use Simpson's rule to obtain a value of $\displaystyle\int_0^1 \frac{dx}{1+x^2}$ working to 5 decimal places.

Use the exact result $\left(\dfrac{\pi}{4}\right)$ to deduce an approximation to π.

(M.E.I.)

E.4. Derive Simpson's rule for approximate integration, viz:

$$\int_a^{a+2h} f(x)\, dx = \frac{h}{3}\{f(a) + 4f(a+h) + f(a+2h)\}$$

Extend the rule to express $\displaystyle\int_a^b f(x)\, dx$ in terms of an odd number of equally spaced values of $f(x)$.

Using 5 ordinates calculate the approximate value of $\displaystyle\int_0^1 \frac{dx}{\sqrt{(1+x^3)}}$ as closely as your 5-figure tables permit.

(Lond.T.Cert.)

E.5. [**Note** that a machine need not be used for this question].

Tabulate the function $f(x) = \sqrt{(\tfrac{1}{2}x + x^2)}$ for $x = 1 \cdot 0,\ 1 \cdot 1,\ 1 \cdot 2,\ 1 \cdot 3$ and $1 \cdot 4$, using square roots to 4 significant figures.

Show that to the accuracy obtained the function is indistinguishable over this domain from a function of the form $p + q(x-1)$ where p and q are constants. Deduce a short method of evaluating $\displaystyle\int_a^b f(x)\, dx$ where $1 \leqslant a < b \leqslant 1 \cdot 4$ and hence or otherwise evaluate $\displaystyle\int_{1 \cdot 02}^{1 \cdot 32} f(x)\, dx$ to 3 significant figures.

(M.E.I.)

E.6. The following purports to be a table of $f(x) = 1 + x + x^3$ and of $\sqrt{f(x)}$. Check the table for suspected errors, correct any, and evaluate $\int_0^{0\cdot8} \sqrt{f(x)}\, dx$ by Simpson's rule. Estimate the reliability of your result using the fact that halving the interval reduces the error of the Simpson approximation to about $\frac{1}{16}$.

x	f(x)	$\sqrt{f(x)}$
0	1	1
0·1	1·101 00	1·048 81
0·2	1·208 00	1·099 09
0·3	1·327 00	1·151 95
0·4	1·464 00	1·209 96
0·5	1·625 00	1·274 75
0·6	1·861 00	1·364 18
0·7	2·043 00	1·429 33
0·8	2·312 00	1·520 52

E.7.

(i) The approximate value of $\int_{-1}^{1} f(x)\, dx$ is required for a variety of functions of **x**. Show how to apply (a) the trapezium rule, (b) Simpson's rule if in each case only *three* ordinates are to be used. Under what conditions will each give the exact value?

(ii) Using Simpson's rule with 5 ordinates evaluate $\int_1^2 \sqrt{\sin x}\, dx$ as accurately as your tables permit. (Lond.T.Cert.)

E.8.

(i) Tabulate values of $f(x) = \sqrt{\{64 + (x-4)^3\}}$ for integral values of **x** from **0** to **8** inclusive and sketch the graph of $f(x)$ over this domain.

(ii) Given that $F(X) = \int_0^X f(x)\, dx$, use Simpson's rule and your tabulated values of $f(x)$ to compute $F(2)$, $F(4)$ and $F(6)$. (M.E.I.)

E.9. It is required to tabulate the function 2^x for $x = 0(0\cdot125)1$ to 6D. Obtain the values for $x = \frac{1}{2}, \frac{1}{4}, \frac{1}{8}$ by successive square-root procedure; the remaining values may be deduced and checked by squaring.

Evaluate $\int_0^1 2^x\, dx$ and use the fact that the true value is the reciprocal of $\log_e 2$ to deduce a value for this number.

Exercise F. Miscellaneous harder questions (e.g. M.E.I. S papers)
F.1. The table shows the resistance of a piece of metal wire at three
different temperatures:

θ	0	100	444·2
R	13·05	18·42	41·10

(i) Fit a linear equation $R = A\theta + B$ to the first two values.

(ii) Determine **k** so that the equation $R = A\theta + B + k\dfrac{\theta}{100}\left(\dfrac{\theta}{100} - 1\right)$
fits all the given values.

(iii) Show how, given a value of **R,** the value of θ may be deduced
by using as a first approximation the value from the linear
equation. Find the value of θ for $R = 25·00$.

F.2. Show that if tabular values are given for $x = 0, 1, 2, 3$ of the function

$$f(x) \equiv a + bx + cx(x-1) + dx(x-1)(x-2)$$

then the values of **b, c, d** can be obtained from the differences derived
from the table, e.g. **b** from $f(1) - f(0)$.

Given the following table for a cubic polynomial $g(x)$, find and check
the values of **a, b, c,** and **d** required to express it in the form given
above; and hence or otherwise evaluate $g(5), g(6), g(7)$ and $g(8)$.

x	0	1	2	3	4
g(x)	87·25	89·37	91·49	96·85	108·69

(M.E.I.)

F.3. A step-by-step solution of a certain differential equation involves
finding y_1, y_2 and y_3 given that

$$y_{n+2} - 2·000625 y_{n+1} + y_n = 0·625$$

and the terminal values $y_0 = 0$, $y_4 = 5·0045$.

By solving the three simultaneous equations obtain the required
values to 4 decimal places.

F.4. The following values of $f(x)$ are known:

$$f(1) = 10, \ f(3) = 4, \ f(4) = 1, \ f(7) = -0·8$$

Assuming $f(x)$ to be the polynomial of lowest degree consistent with
the data find $f(2), f(5), f(6)$.

Show by a difference table that your solutions are consistent with any assumption you have made.　　　　　　　　(Lond.T.Cert.)

[*Hints.* One method is to write $f(2)=k$ and form a difference table which is then made consistent. A more subtle method uses the remainder theorem:

$$f(x) \equiv (x-1)\ Q(x) + 10$$

Then determine the coefficients of $Q(x) \equiv a(x-3)^2 + b(x-3) + c$].

F.5.

　(i) When is a set of simultaneous linear equations said to be ill-conditioned ? Devise one numerical example of an ill-conditioned set of two equations and one example of a well-conditioned set of two equations.

　(ii) A function $f(x)$ which satisfies the differential equation $f''(x) = 4f(x)$ is known to have the following values:

$$f(0\cdot15) = 1\cdot045\ 34$$
$$f(0\cdot20) = 1\cdot081\ 07$$
$$f(0\cdot25) = 1\cdot127\ 63$$

　　Test the accuracy of the approximate formula for $f''(x)$ in terms of $f(x-h)$, $f(x)$ and $f(x+h)$ by evaluating $f(0\cdot25)$ from the other data. Carry out this step-process to obtain values of $f(0\cdot05)$, $f(0\cdot10)$, $f(0\cdot30)$, and $f(0\cdot35)$. Estimate the maximum errors in the extreme values assuming them to be entirely due to rounding errors in the data.　　　　　　　　(M.E.I.)

F.6. Transform the differential equation:

$$y'' + 2y' + y = 0$$

into a difference equation, making use of the approximations:

$$h^2(y'')_0 \approx y_{-1} - 2y_0 + y_1$$
$$2h(y')_0 \approx -y_{-1} \quad\ + y_1$$

where y_{-1}, y_0, y_1 are the ordinates at $x-h$, x and $x+h$, respectively.

　Given that $y=1$ at $x=0$ and $y=0\cdot4$ at $x=0\cdot8$, estimate y at the points $0\cdot2$, $0\cdot4$ and $0\cdot6$ to 2 decimal places.　　　　　　　　(M.E.I.)

F.7.

　(i) Show that if a function $f(x)$ can be expressed over the domain $(a \leqslant x \leqslant b)$ by a cubic polynomial in x then within the domain

$(a+h \leqslant x \leqslant b-h)$ the expressions:

$$h^2 f''(x) \quad \text{and} \quad f(x+h) - 2f(x) + f(x-h)$$

are identical.

(ii) The function $g(x)$ satisfies the equation:

$$g''(x) = \sin(\pi x) - g(x)$$

and it is given that $g(-0 \cdot 1) = -0 \cdot 26466$ and $g(0) = 0$. Assuming that $g(x)$ satisfies the conditions for $f(x)$ described in (i), express $g(x+h)$ in terms of $g(x)$ and $g(x-h)$, and hence evaluate the function step by step for $x = 0 \cdot 1, 0 \cdot 2$ and $0 \cdot 3$, retaining 5 decimal places throughout the work. (M.E.I.)

F.8. In a probability problem it is required to find to 3 significant figures the value of $\dfrac{b}{a}$ where:

$$a = \int_0^{0 \cdot 8} \cos^{-1} x \, dx \quad \text{and} \quad b = \int_0^{0 \cdot 8} x \cos^{-1} x \, dx.$$

The integrals may be evaluated by Simpson's rule using 5 equally spaced ordinates.

[*Note* that $\cos^{-1} x$ implies a radian value, but since a ratio is being found you may if desired work with $\cos^{-1} x$ in degrees.]

F.9. (a) Show that if $0 < t < 1$, the error in equating $1/(1+t)$ to the sum of the first n terms of the expansion $1 - t + t^2 - \cdots + (-1)^r t^r + \cdots$ is less than the $(n+1)$th term. Deduce, explaining your reasoning carefully, a similar result for:

$$\ln(1+t) \approx t - \frac{t^2}{2} + \cdots + \frac{(-1)^r t^r}{r} + \cdots \qquad [\ln = \log_e]$$

(b) A computer incorporates a program for calculating the natural logarithm of any positive number. The first stage is to express the given number x as $b^n y$, where b is a fixed number between 1 and $1 \cdot 5$ with known natural logarithm, and $1 \leqslant y < b$. The second stage is to calculate $\ln y$ by the expression given above.

Either write a flow chart for finding n and y for given positive x.

Or given that $\ln(1 \cdot 284\ 025) = 0 \cdot 250\ 000\ 0$ use the method described to find $\ln(1 \cdot 300\ 000)$ to 6 decimal places.

[The reciprocal of $1 \cdot 284\ 025$ is $0 \cdot 778\ 801\ 0$.]

F.10. The velocity of a body is given in terms of its position by the equation $(dx/dt)^2 = a - x^2 - bx^3$, where a and b are experimentally determined to be **1·08** and **0·070**, respectively. Evaluate dt/dx for $x=0$, **0·2, 0·4, 0·6** and **0·8**; and given that $t=0$ when $x=0$ find by numerical integration the values of t when $x=0·4$ and **0·8**, giving your results to 3 decimal places.

State carefully how you would estimate the value of t when $x=1$, explaining why it is difficult to get a reliable value. (M.E.I.)

F.11. (a) It is known that Simpson's rule with three ordinates is accurate when applied to a cubic polynomial. A certain function $F(x)$ can be expressed, over the domain $-c \leqslant x \leqslant c$, as:

$$F(x) \equiv b_0 + b_1 x + b_2 x^2 + b_3 x^3 + g(x)$$

where $0 \leqslant g(x) \leqslant kx^4$, k being a given positive constant. Show that there is an upper bound of $\frac{2}{3}kc^5$ for the absolute error in using Simpson's rule with three ordinates to evaluate:

$$\int_{-c}^{c} F(x) \, dx$$

(b) Tabulate the function $F(x) \equiv (1+x)^{-1}$ for $-0·2(0·1)0·2$ to 7 decimal places. Use Simpson's rule with three ordinates to evaluate the integral for the cases $c=0·1$ and $c=0·2$, and calculate the bounds of error as in (a) with the help of the identity:

$$\frac{1}{1+x} = 1 - x + x^2 - x^3 + \frac{x^4}{1+x}$$

Compare the predicted bounds with the actual errors, given the following values of the function $L(z) = \ln\left(\frac{1+z}{1-z}\right)$ to 7 decimal places:

$$L(0·1) = 0·200\ 670\ 7$$
$$L(0·2) = 0·405\ 465\ 0 \qquad \text{(M.E.I.)}$$

APPENDIX 1

PROGRAMMING AND THE COMPUTER

1. In these exercises the art of *programming* a computation has been developed. It is not sufficient to have a method which is sound; broken down into 'units of process' which the machine can handle, it should also
  (i) proceed smoothly and economically;
  (ii) lend itself to a clear layout which makes errors less likely;
and (iii) be capable, if long, of intermediate checking, to avoid wasted labour. (Study the example in 6.2.)

In non-iterative processes, e.g. solution of simultaneous equations by Gauss' elimination method, a checking method is built into the process itself.

Iterative methods are self-checking; and the worst that can happen after, say, a copying or re-entering error is that more iterations are required before the result is achieved.

It will be noticed that the programs which lend themselves to flow-diagram representation are all iterative. The instructions which form part of a loop are in fact repeated, with modifications which are provided for, until a certain routine test is satisfied and 'directs' the process out of the loop. Look again at the diagram on p. 22 from this point of view.

Electric desk-machines

An electrically-driven machine which provides for automatic division is in fact designed according to the Figure on p. 22. No bell actually rings, but a close scrutiny will show that every subtraction actually continues until zero is passed, and the last backward turn is then retraced.

Some electrical machines will only multiply digit by digit: pressing the '4' key, for example, simply causes the crank to turn 4 times. On fully automatic machines an extra register holds the multiplier digits, preselected: the '×' key then initiates multiplication as an iterative process, continued digit by digit so long as there are non-zero digits shown in this register.

These machines are not in any true sense computers; they simply carry out one, or at most two, planned patterns of calculation. They do however suggest to the mind the desirability of machines which could carry out not only *more* flow-patterns but preferably *any required type* of pattern, according to a programme of instructions fed into it in advance.

Around 1840, Charles Babbage worked out all the main principles of such a machine, to be fed with instructions on punched cards, and the partly constructed machine (purely mechanical) is to be seen at the Science Museum, London. It is possible today for a young student who has some experience of calculation with a desk-machine to list for himself some of the basic features he would wish such a machine to have, if the intervention of the operator is not to be required during the program.

Requirements for a Computer

Babbage called the part of his machine where numbers are currently being processed, the MILL. It was required (like the automatic desk-machine of today) to add, subtract, multiply or divide pre-set numbers. We also require it to *compare* two numbers: the test for leaving a loop is commonly dependent on such a comparison (e.g. on p. 24, 'Is T negligible?' in precise terms might be 'Is $|T| < 10^{-6}$?').

In place of our pencil-and-paper we need STORAGE facilities; for data, for intermediate and final results, and (not least) for the actual program itself and any ancillaries associated with it.

Above all, to obviate the interference of the operator at each new phase of the program we need a CONTROL UNIT. Two factors have contributed jointly to the success of this, the characteristic feature of the computer, viz: (1) algebraic logic, initiated by George Boole in 1840, and (2) the invention of electronic pulse-operated devices.

The relation between the parts of the computer and its input and output units can be sketched thus:

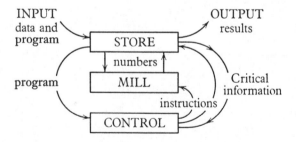

One can use a computer successfully with only slight knowledge of the inside processes, by storing in the computer a suitable 'translator program'. This allows the programmer to feed into the machine a set of coded instructions not far removed in essence from a flow diagram. While being read in, these are checked for translatability into machine process; and when the whole program is stored it awaits the final button to set it into operation. (Compare how, in an electric machine, the feeding in of a multiplier does not itself cause the machine to multiply.)

The student who wishes to get an introduction to computer programming will be well advised to go to a short course at a college which has its own computer or computer terminal. Those aiming to do professional work in the computer field will do well to start on one of the high-level languages as soon as possible. If however the object is to *teach* the elements of the subject (which has become an essential feature of a broad modern education) then one should see what can be done with a simpler language such as BASIC or TELCOMP. These can be used effectively after a very short introduction, especially if the computer embodies a diagnostic program, i.e. if it indicates at once to the user whenever a linguistic error is committed.

APPENDIX 2

INSTRUCTORS' NOTES

STANDING ORDERS: for students using the machine.

(1) See that the machine is safely placed, i.e. on a firm flat table or desk.

(2) See that you are in no danger of being jogged while at work: no other person should be overlooking your work unless he is
either teaching you
or being taught by you.

(3) Always complete one movement before starting another.

(4) **NEVER FORCE ANY LEVER:** in case of jamming — put machine away and report the difficulty.
(In many instances what appears to be a jam is only a stoppage due to *either* failure to complete a turn *or* failure to effect a full clearance. The remedy in each case is obvious.)

(5) After use: clear all registers

return carriage and
locking levers } to normal,

and put the machine away unless directed otherwise.

Further notes for class-instruction

If you use a machine for class-teaching with junior students it is advisable to rehearse *in detail* the calculations proposed.

The students will naturally be anxious to carry out manual operations themselves, and it is tempting to devise a cooperative calculation such as adding the ages of the students in months, letting each one add his own. In fact such a project is very liable to go wrong, unless a previously briefed student is standing over the machine checking that every result is entered correctly. Individual projects (e.g. for each to turn his age in years and months into months) are better because (i) one operator's error doesn't affect a final result and (ii) the worker will check the result on paper and will be able to compare his speeds with and without the machine. It is *still* a good idea to have a student supervisor to minimise error. With a lock it may be practicable to set the same figures in the locked part, so that a running total also is obtained.

APPENDIX 3

MACHINES

A number of companies distributing in the U.K. have hand-machines with full tens-transmission (*essential*) and back-transfer (*desirable*) at educational prices in 1965 from £35 up to £58. These all have lever setting, keyboards being more expensive. Their capacity is usually 10 (set) by 8 (cycle) by 13 (accu.). Choice will depend not only on cost, of purchase or hire or hire-purchase, but also on the service facilities available in your area.

Some suitable models are listed below; the order is alphabetical (of distributors' names):

Multo: Addo Ltd., 47 Worship St., London E.C.2.
Teacher II: ADM Ltd., 64 King St., W.6.
Odhner: Block and Anderson, Cambridge Grove, W. 6.
Nippon: Broughton's, 6 Priory Rd., Bristol 8.
Mentor: Muldivo Ltd., 28 Banner St., E.C.1.
Brunsviga 13: Olympia Ltd., 203 Marylebone Rd., N.W.1.

Keyboard (*including electrical*) *machines* are sold by some of the above and also by:

Monroe Calc. Mach. Co., Litton Ho., 27 Goswell Rd., E.C.1. and Olivetti Ltd., 30 Berkeley Sq., W.1.

A pocket machine 8 × 6 × 11 (no transfer) is the *Curta I:* London Office Machines, 5 Lower Belgrave St., S.W. 1.

(All addresses are in London unless otherwise stated.)

APPENDIX 4

BOOKS OF FUNCTION TABLES

1. Elementary

If 4-figure tables are used, then those of Miller and Powell, *The Cambridge Mathematical Tables* (C.U.P.), are recommended.

Generally, the machine operator will feel 4-figure tables of functions to be inadequate. This is not true in statistics, in which the table requirements are exacting in a different way.

2. Reference

No machine room is complete without six-figure tables for reference, preferably *Chambers' Shorter 6-figure tables* (Ed. Comrie). These have interlinear differences, and the arguments are close enough for linear interpolation to be used throughout (with a table of proportional parts to aid the calculation).

3. Portable, for everyday use

The norm must be regarded as 5 figures for the machine user. His work is however largely misdirected if based on unreliable 5-figure tables e.g. a table built in the traditional 4-figure style *with columns of mean differences.* This type of table will undoubtedly give way in favour of the clear but slightly more bulky layout shown in the forthcoming Miller and Powell *Advanced Mathematical Tables* (C.U.P.).

A table which combines the older layout with *true* differences is Attwood *Advanced Mathematical Tables* (Macmillan). A student equipped with one of these (and any extra statistical tables required for his work) and with access to a hand-machine, is well equipped to tackle all the common types of numerical task. Needless to say he may well move over to a computer terminal when the shape of the process has become clear. It must however be emphasised that omitting the machine overture is responsible for much waste of expensive computer time.

SOLUTION NOTES

1.8 First difference $= a(3n^2 + 3n + 1) + b(2n + 1) + c$;
second $= a(6n + 3) + 2b$; third $= 6a$; fourth $= 0$.

3.1 (a) (i) 37·28, 37·3, (ii) 4·210, 4·21
3.4 (i) n.mile total $\times 0·002 = 0·20$ km, to 2D.
(ii) max. error 0·25 n.miles $= 0·45$ km.
3.5 (a) 1721, 1·721, (b) 260·8 written 260·800 as 1, ml.
3.6 (a) 237·2 min, 3·95 h, (b) 57° 17·6′
(c) 36·87°, 0·6435 radians, (d) 193·1 n.mile, 357·6 km.

3.7 (i) $0·895 \pm 0·009$
(ii) $5·25 \pm 0·01$ since x, y uncertain to 0·005
(iii) $x\,\Delta y + y\,\Delta x = 0·005(x + y) = 0·026$ to 2S
$\therefore$ say $xy = 3·300 \pm 0·026$
(iv) $2xy\,\Delta x + x^2\,\Delta y = 0·005x(x + 2y) = 0·135$
giving $x^2 y = 14·916 \pm 0·135$
Extreme values directly: 15·050, 14·779

4.1 $S_n = 1 + r + r^2 + \ldots + r^{n-1}$ (n terms) and $S_{n+1} = S_n + r^n$; but a more useful relation is $S_{n+1} = 1 + rS_n$, which gives a simple programme of calculation starting with $S_0 = 0$ or $S_1 = 1$.

4.3 $30^2 - 29^2 = (30 - 29)(30 + 29) = 30 + 29$
and so on: thus we are summing the integers 30, 29…21.

4.6 (a) $\Delta f_r(n) = \dfrac{(n+1)^{[r]}}{r!} - \dfrac{n^{[r]}}{r!} = \dfrac{(n+1)^{[r-1]}}{r!} \left\{ (n+r) - n \right\} = \dfrac{(n+1)^{[r-1]}}{(r-1)!}$

4.8 (i) $\Delta a = 0·005$ for every a, and thus
error $\approx 0·005(x^4 + x^3 + \ldots + 1) = 0·005 \times 8·21$
$= 0·041$ to 2S
(ii) error in $x^4 \approx 4x^3\,\Delta x$, and so on.
Calculate $(4a_0 x^3 + 3a_1 x^2 + 2a_2 x + a_3)\,\Delta x$
or recalculate $f(x)$ for $x = 1·26$.

4.10 The settled value of 4th difference seems to be 24, and the errors presumed in the other values ($+9$, -36, $+54$, -36, $+9$) indicate an error $+9$ at the 'centre of disturbance'. The reading 6·10 should be 6·01: transposition of figures in copying was a likely cause.

5.1 (b) The rounding error for 4S is anything from 0 to 5 ($+$ or $-$) in the fifth figure. Taking x as exact, $x \times (1/x$ rounded) could be in error by as much as $1{\cdot}249 \times 5 \times 10^{-5}$ for the first result and $0{\cdot}08334 \times 5 \times 10^{-3}$ for the second: thus rounding is potentially seven times as great a source of error in the second case as in the first.

5.2 Max. error in numerator xy due to rounding
$$\approx x\,\Delta y + y\,\Delta x = 39{\cdot}2 \times 0{\cdot}005 + 2{\cdot}47 \times 0{\cdot}05$$
$$= 0{\cdot}320$$
$\therefore$ in $xy/z \approx 0{\cdot}320/43 = 0{\cdot}007$

Cf. greatest and least values by direct calculation.

5.3 (ii) A general result for $z = kx^m y^n$ which is often quoted in scientific work is shown as follows:

We have $\log_e z = \log_e k + m \log_e x + n \log_e y$.

Then if x alone suffers a small change Δx, we can use the result
$$\frac{1}{z}\frac{dz}{dx} = \frac{m}{x}$$

to obtain
$$\frac{\Delta z}{z} \approx \frac{m\,\Delta x}{x}$$

[Fractional or $\%$ error in z, due to x alone, is $\mid m \mid$ times the fractional or $\%$ error in x.]

A similar result is true for y alone; and if we neglect their interaction we obtain for the greatest numerical error in z the equation
$$\max \left| \frac{\Delta z}{z} \right| \approx \left| m\,\frac{\Delta x}{x} \right| + \left| n\,\frac{\Delta y}{y} \right|$$

We have seen this in the special cases $z = xy$, $z = x/y$.

6.1 It is worth while to consider whether build-up division gives an easier programme. [Set x; accumulator clear and fully to right; crank and shift so as to get N in accumulator correct to as many figures as possible.] In either case the flow diagram for $\sqrt{N}$ is as follows:

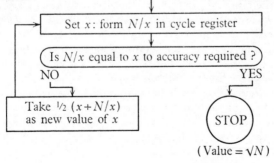

(Value $= \sqrt{N}$)

This flow diagram can be given more precision by stating (progressively) the number of significant figures to retain at each stage.

7.9 Yes. Although the second trial by itself gives a smaller value of χ^2, the two regarded as a *single longer trial* gives a much larger value; the persistently low frequencies 1 and 6 contribute largely to this, and the result accords with a common-sense view.

8.2 (b) Any lingering doubts about the sources of error can be dispelled by an 'elementary arithmetic' solution with exact multipliers 14 and 17, giving $x = 3\frac{2}{9}, y = 9\frac{1}{9}$.

Graphically we see that the lines are nearly parallel, so that a slight change has a large effect on the point of intersection.

Algebraically, the determinant $\begin{vmatrix} a_{11} & a_{12} \\ a_{21} & a_{22} \end{vmatrix}$ is small compared with the actual values of the a's.

8.8 (i) $-5x + 6y - z + 1 = 0$, etc.

10.4 $\dfrac{dq}{dT}$ requires knowledge of $\delta^3 q$ and is therefore only obtainable over the range stated. Even so, its acceptance is scarcely 3S (below 32·5) or 2S (above); and the same is true of $f(T)$. We do, however, obtain a steady value of $\Delta^3 f$ for this range and to this accuracy: thus f can be expressed well enough by a cubic polynomial, either in terms of T or (as the chemist prefers) in terms of absolute temperature $T + 273$, written below as θ.

The above work indicated the need for a greater range of experimental values and also an improved numerical method. All this has now been achieved, as below.

q was fitted to a formula of this type:

$$a_0/\theta + a_1 \log_e \theta + a_2 + a_3 \theta$$

(This corresponds to a *quadratic* polynomial for f)

The method was (i) to compute $g = \Delta^2(\theta q)$ and $h = \Delta^2(\theta \log_e \theta)$ over the range; (ii) to find, graphically, the value of a_1 such that $g - a_1 h$ is sensibly constant; (iii) knowing that $\theta q - a_1 \theta \log_e \theta$ is a quadratic function, viz: $a_0 + a_2 \theta + a_3 \theta^2$, fit the values of these coefficients to three temperatures spread over the range of values; (iv) check the fit of the q formula over the whole range, and hence confirm the formula for $f = -a_0 + a_1 \theta + a_3 \theta^2$.

This formula agreed better with results obtained in other ways than did a cubic of slightly closer fit. This indicates that one would have been justified in smoothing the experimental data more than was actually done.

ANSWERS TO REVISION EXERCISES

A.1 0·789 53, 0·917 44, 0·000 04, 0·000 00

A.2 0·336 65, 0·000 00, 1×10^{-5}

A.3 2·7791, $-0·4632$, $-1·1164$, 1·8528

A.4 0·017 324, 0·022 573 (or 4)

A.5 1·10, $-1·25$, 1·83

A.6 0·9107, 0·1982, $-0·3684$, $-0·0003$, $-0·0001$, 0·00030

B.1 (a) $y = 72$ at $x = 6$ (b) 0·3473

B.2 2·13, 2·166

B.3 0·944

B.4 For 0·865 read 0·856. For 0·991 read 0·919. Value 0·887. D^2 positive, curve concave upwards, value 0·887 too large.

B.5 8, entry 1·76 in $g(x)$ should be 1·67.

B.6 133·04 should be 133·40, an error which easily occurs in transcribing.
$a = 103·32$, $b = 7·30$, $c = 0·06$, $d = 0·0067$.
121·05 from cubic; 121·07 by interpolation.

B.7 $-0·18$ above, 1·30 above, 2·54 below.

B.8 0·3222, i.e., 0·32 to 2S.

B.9 $-0·8393$, i.e., $-0·84$ to 2D.

B.10 0·759 836, 0·438 691, product compared with value $\frac{1}{3}$.

B.11 1·866(1)

B.12 $f(x) = N - \dfrac{1}{x^2}$ gives $x_{n+1} = \frac{1}{2}x_n (3 - Nx_n{}^2)$, 0·577 350

C.1 0·40551 using 5-figure tables

C.2 2·152

C.3 1·276

C.4 1·32

C.5 1·261

C.6 5·357, 3·961, 2·996, 2·280, 1·702
First plot $y = (1 + \ln x)/x$ for $x = 1(1)6$.

C.7 2·048

C.8 (a) For 1·5057 read 1·5076
(b) $c = 0·325 \Rightarrow x = 0·77$ and $y = 1·77$

D.1 4·781 or 4·782. Is $|x - y| < 5 \times 10^{-5}$?

D.2 $x^2 + mx + n = 0$
(2, 0·5) (2, −0·5) 3·980, 0·0204
Numerically smaller root $= n/$(larger root)

D.3 $a := \dfrac{\pi}{2}$ ⟨Is $k = 0$?⟩ YES→Print a

NO → Loop as before

$\dfrac{\pi}{32}$, $\dfrac{\pi}{32}$

D.5 **(a)** 000110⎫   111100⎫   000000⎫
      010110⎭   101101⎭   000000⎭

(b) To get numbers *out* of the accumulator, instruction *Add* must be used with zero.

Set (1), *SL*, Add (2, 4):8n into accumulator.
Add (4, 3): sum 10n into location 3 as required.

E.1 0·477 25 0·954 50: second difference changes sign.
E.2 1·007 63 (both). Possibly correct to 5D, certainly 4D.
E.3 0·78539
E.4 0·9097
E.5 0·419 using either trapezium or mid-ordinate rule.
E.6 $f(0·6) = 1·816\ 00$ and $\sqrt{f(0·6)} = 1·347\ 60$
Simpson mean of all values is 1·227 22, of five values 1·227 27.
If difference $= 15 \times$ error, then first result is probably correct to all figures; and certainly the integral $= 0·981\ 78$ to 5S.
E.7 0·9777
E.8 (ii) 10·60 (or 10·61), 26·35, 42·59
E.9 0·693 147

F.1 0·0537, 13·05, 0·2745
$\theta \approx 222·5$ converges to 210·6
F.2 130·25, 164·77, 215·49, 285·65
F.3 0·3126, 1·2505, 2·8141
F.4 7·2, −1·2, −2·0
F.5 (The function is ch 2**x**). Maximum error 0·000025
F.6 0·80 (0·796), 0·63(3), 0·50(3)
F.7 0·264 66, 0·529 76, 0·795 44
F.8 0·347
F.9 **(b)** 0·262 364
F.10 0·395, 0·887, $\approx 1·56$
F.11 **(b)** 0·20067 34   0·405 471 4
Errors $+2·7 \times 10^{-6}$ (bound $7·4 \times 10^{-6}$)
     $+6·4 \times 10^{-6}$ (bound $2·7 \times 10^{-4}$)

BIBLIOGRAPHY

(In each *section*, titles are in approximate order of difficulty)

Tables (see also Appendix IV):
  Miller and Powell: *Elementary Mathematical Tables* (Cambridge)
                    Advanced Tables (Cambridge) in preparation
  Ed. (Comrie): *Shorter 6-figure Mathematical Tables* (Chambers)
  Interpolation and Allied Tables (H.M. Stationery Office)

Machines (teaching method and exercises):
  French: *Introduction to Calculating Machines for Schools* (Macmillan)
  Behmber and Jewell: *Learning to use the Desk Calculator*: a set of
  practice cards for class use in schools, with handbook (Macmillan)

Numerical Analysis (general):
  Morton: *Numerical Approximation* (Routledge)
  Wooldridge: *Introduction to Computing* (Oxford)
  Wilkes: *Introduction to Numerical Analysis* (Cambridge)
  Modern Computing Methods: (H.M. Stationery Office)

Computers (for young students):
  Lovis: *Computers* (I and II) (Arnold)

Computers (and programming):
  Phillips and Taylor: *Computers* (Methuen)
  Corlett and Tinsley: *Practical Programming* (Cambridge)
  Hollingdale: *High Speed Computing* (E.U.P.)

Special Topics (in alphabetical order):
  (Finite differences); Freeman: *Finite Differences* (Cambridge)
  (Flow diagrams); Moakes: *Pattern and Power of Mathematics*, Bk. 6
  (Macmillan); and extended to simple computing: ditto, Bk. 7
  (Linear equations); Cohn: *Linear equations* (Routledge)
  (Mathematical basis); Moakes: *Core of Mathematics* (Macmillan)

  (Matrices):
  Matthews: *Matrices* (I and II) (Arnold)
  Brand and Sherlock: *Matrices, Pure and Applied* (Arnold)
  Neill and Moakes: *Vectors, Matrices and Linear Equations* (Oliver
  and Boyd)

  (Statistics):
  Brookes and Dick: *Introduction to Statistical Methods* (Heinemann)
  Paradine and Rivett: *Statistical Methods for Technologists* (E.U.P.)

INDEX